Z18-36

M.C. Goodall

University of Alabama.

205-934-5657

Grundlehren der mathematischen Wissenschaften 231

A Series of Comprehensive Studies in Mathematics

Serge Lang

Elliptic Curves Diophantine Analysis

Springer-Verlag
Berlin Heidelberg New York 1978

Serge Lang
Department of Mathematics, Yale University,
New Haven, CT 06520, U.S.A.

AMS Subject Classification (1970): 10 B 45, 10 F 99, 14 G 25, 14 H 25

ISBN 3-540-08489-4 Springer-Verlag Berlin Heidelberg New York
ISBN 0-387-08489-4 Springer-Verlag New York Heidelberg Berlin

Library of Congress Cataloging in Publication Data. Lang, Serge, 1927–. Elliptic curves (Grundlehren der mathematischen Wissenschaften: 231). Bibliography: p. Includes index. 1. Diophantine analysis. 2. Curves, Elliptic. I. Title. II. Series: Die Grundlehren der mathematischen Wissenschaften in Einzeldarstellungen; 231. QA242.L234. 512'.74. 77-21139.

Printed in Germany

Typesetting: William Clowes & Sons Limited, London, Beccles and Colchester.
Printing and binding: Konrad Triltsch, Würzburg
2141/3140-543210

Foreword

It is possible to write endlessly on elliptic curves. (This is not a threat.) We deal here with diophantine problems, and we lay the foundations, especially for the theory of integral points. We review briefly the analytic theory of the Weierstrass function, and then deal with the arithmetic aspects of the addition formula, over complete fields and over number fields, giving rise to the theory of the height and its quadraticity. We apply this to integral points, covering the inequalities of diophantine approximation both on the multiplicative group and on the elliptic curve directly. Thus the book splits naturally in two parts.

The first part deals with the ordinary arithmetic of the elliptic curve: The transcendental parametrization, the p-adic parametrization, points of finite order and the group of rational points, and the reduction of certain diophantine problems by the theory of heights to diophantine inequalities involving logarithms. The second part deals with the proofs of selected inequalities, at least strong enough to obtain the finiteness of integral points.

The historical development is such that the first part represents a relatively mature state of the subject, whereas the second part is in a state of flux (due in large measure to the Baker method), so that no current account can be regarded as in any way definitive. The selection of which theorems and which methods to include was based on emphasizing the analogy between operations on the multiplicative group and operations on the elliptic curve, and was meant to give typical results, for instance the first inequality of Baker–Feldman, whose proof is less involved than some others, and is similar to the proof in the subsequent chapter working directly with the elliptic logarithms. The last two chapters illustrate two methods of descent (due to Baker, with some improvements from Cijsouw–Waldschmidt, Tijdeman and Van der Poorten). At the moment they have no analogue in the elliptic case, but it seemed important to make available to the reader as many methods as possible. Finally, the theorem given there (Baker–Tijdeman) leads to the Catalan problem, and it would be interesting to have analogous formulations for elliptic curves.

Elliptic curves serve as a prototype for abelian varieties, as a special case of curves, and as a means of handling other curves by the theory of correspondences. Using concrete formulas, one can get into the theory of elliptic curves without much mathematical background, and one can reach rapidly substantial levels of depth.

However, it should not be forgotten that curves of higher genus ultimately require a thorough understanding of their Jacobians, which cannot avoid the tools developed by algebraic geometers in the last 30 years. Mazur's success in proving that some simple factor of the Jacobian of modular curves over the rationals has only a finite number of rational points is a testimony to the power of the most general tools provided by algebraic geometry. Via this higher dimensional theorem, one obtains bounds for torsion points on elliptic curves over the rationals which so far have not been obtained by other methods. Thus even as the elliptic curves affect the diophantine properties of other curves, conversely curves of higher genus like the modular curves, or the Fermat curve as in Demyanenko or Kubert-Lang affect the diophantine properties of elliptic curves.

Parallel to the pure arithmetic theory over number fields lies the algebraic-geometric theory of algebraic systems (over the complex numbers if you wish), where sections play the role of rational points. This is the point of view taken in *Diophantine Geometry*. Its origin lay in Severi's recognition of the connection between the "theorem of the base" (finite generation of the group of divisor classes modulo algebraic equivalence) and the Mordell–Weil theorem (finite generation of the group of rational points on an abelian variety). The theorem of the base was proved by Néron in his thesis, and a closer tie between the two theorems was established by Lang–Néron, formulating the relative Mordell–Weil theorem for algebraic families of abelian varieties (the group of sections modulo constant sections is finitely generated). In dimension 1, I showed how the presence of infinitely many integral sections in a family of affine curves of genus $\geqq 1$ implies that the family splits, and almost all sections come from constant ones. My conjecture (transposing Mordell's) that for genus $\geqq 2$ this should also apply to rational sections was proved by Manin. Another proof was subsequently given by Grauert. Both proofs lead, in different ways, to differential geometric considerations on the fiber space. Shafarevič's theorem that there is only a finite number of elliptic curves (up to isomorphism) over a number field, having good reduction outside a given finite set of primes led to Paršin's results along these lines for algebraic families of curves. Néron's classification of minimal models of elliptic curves over discrete valuation rings parallels Kodaira's classification of pencils of elliptic curves. All of these results would make up a nice new volume of diophantine geometry. (Incidentally, Seligman has observed that the Néron–Kodaira diagrams are dual to the Dynkin diagrams in the theory of Lie algebras. No theoretical reason for this has yet been found.)

The methods used for the problems just mentioned have an essentially algebraic aspect. It is also possible to transport the problems to a setting involving the geometry of several complex variables. Curves of genus $\geqq 2$ may be viewed as the 1-dimensional versions of quotients of bounded domains, or of varieties of "general type". I formulated some analogues of the classical diophantine problems in this context, and some of these have been proved recently, e.g. by Kobayashi and Ochai, the finiteness of the number of surjective meromorphic maps onto varieties of general type; and Mark Green, the hyperbolicity of a subvariety of an abelian variety which does not contain the translation of an abelian subvariety. Conjectures as in [L 5] lead to conjectures concerning algebraic families of abelian varieties, or elliptic curves, over the complex numbers. Assuming for simplicity that the family has no fixed part, let $\sigma_1, \ldots, \sigma_r$ be meromorphic sections, linearly independent over the integers. In each fiber we have a metric corresponding to the torus, and one can define a norm on sections as the sup norm over all points of the parameter variety, in a fixed small neighborhood U of a point. Then the complex analytic analogue of the Baker–Feldman theorem should be that a linear combination of sections satisfies an inequality

$$\|q_1\sigma_1 + \cdots + q_r\sigma_r\|_U \geqq q^{-C},$$

where $q = \max |q_i|$, and C is some constant.

Even as diophantine questions from number theory give rise to problems in geometry (algebraic and differential), conversely number theory can also profit from the techniques of several complex variables (e.g. as introduced by Bombieri–Lang in the theory of transcendental numbers and diophantine approximation, and pursued by Masser, with his theorem asserting that a polynomial having sufficiently many zeros in the unit ball, not too far apart as a function of the degree, must in fact be identically zero).

An advanced monograph like *Diophantine Geometry*, presupposing substantial knowledge in some fields, and thus allowing certain expositions at a level which may be appreciated only by a few, but achieving a certain coherence not otherwise possible, of course does not preclude the writing of elementary monographs. Both coexist amicably. Each achieves different ends. In some sense the present book corresponds to *Diophantine Geometry* on elliptic curves, although of course the theory goes further in the light of progress made in the last 15 years. This is entirely consistent with my conclusion of the review of the first portion of Grothendieck's *Éléments de Géometrie Algébrique* (*Bulletin* AMS, 1961):

> "...If Algebraic Geometry really consists of (at least) 13 chapters, 2,000 pages, all of commutative algebra, then why not just give up? [I was optimistic, it's more like 7,000 pages by now...]
>
> The answer is obvious. On the one hand, to deal with special topics which may be of particular interest only portions of the whole work are necessary, and shortcuts can be taken to arrive faster at specific goals...Projective methods, which have for some geometers a particular attraction of their own, and which are of primary importance in some aspects of geometry, for instance the theory of heights, are of necessity relegated to the background in the local viewpoint of *Elements*, but again may be taken as starting point given a prejudicial approach to certain questions.
>
> But even more important, theorems and conjectures still get discovered and tested on special examples, for instance elliptic curves or cubic forms over the rational numbers. And to handle these, the mathematician needs no great machinery, just elbow grease and imagination to uncover their secrets. Thus as in the past, there is enough stuff lying around to fit everyone's taste. Those whose taste allows them to swallow the *Elements*, however, will be richly rewarded."

On the other hand, the present book is addressed to those whose taste lies with elliptic curves.

Serge Lang

Acknowledgment

I am much indebted to Michel Waldschmidt, Neal Koblitz and David Rohrlich for reading through the manuscript carefully, and for a large number of very useful comments.

I thank Addison Wesley for letting me reprint the first few sections in Chapter I from *Elliptic Functions*, concerning the standard properties of the Weierstrass functions.

I thank the editors of Springer-Verlag for their willingness to share with me the excitement which seems to accompany occasionally the publication of my books. They deserved my acquiescence to their request to eliminate from the foreword statements (not necessarily by me) which might be interpreted as perpetuating unnecessary polemics.

Table of Contents

Part I. General Algebraic Theory

Part I

General Algebraic Theory

Chapter I. Elliptic Functions

§ 1. The Liouville Theorems

By a **lattice** in the complex plane **C** we shall mean a subgroup which is free of dimension 2 over **Z**, and which generates **C** over the reals. If ω_1, ω_2 is a basis of a lattice L over **Z**, then we also write $L = [\omega_1, \omega_2]$. Such a lattice looks like this:

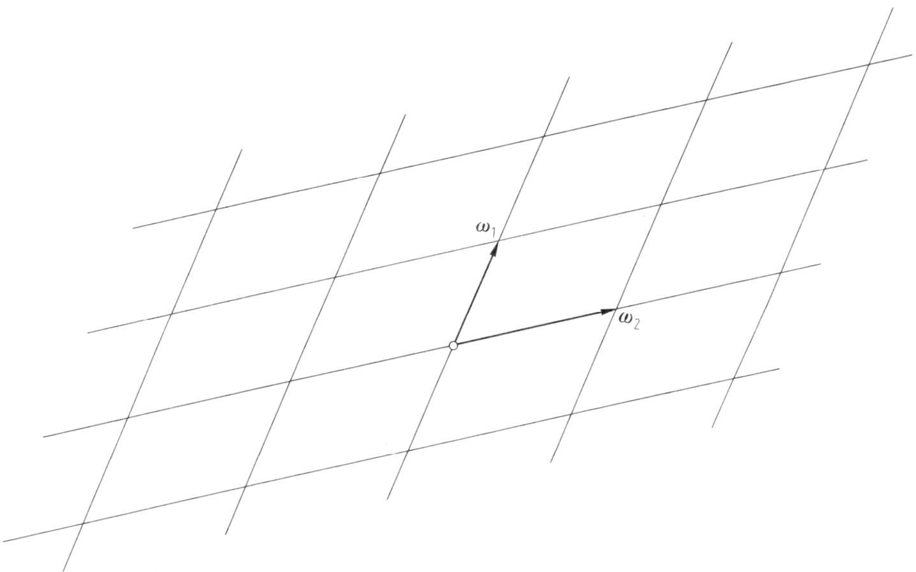

Fig. 1

Unless otherwise specified, we also assume that $\mathrm{Im}(\omega_1/\omega_2) > 0$, i.e. that ω_1/ω_2 lies in the upper half plane $\mathfrak{H} = \{x + iy, y > 0\}$. An **elliptic function** f (with respect to L) is a meromorphic function on **C** which is L-periodic, i.e.

$$f(z + \omega) = f(z)$$

for all $z \in \mathbf{C}$ and $\omega \in L$. Note that f is periodic if and only if

$$f(z + \omega_1) = f(z) = f(z + \omega_2) .$$

An elliptic function which is entire (i.e. without poles) must be constant, because it can be viewed as a continuous function on $\mathbf{C}/L$, which is compact (homeomorphic to a torus), whence the function is bounded, and therefore constant.

If $L = [\omega_1, \omega_2]$, as above, and $\alpha \in \mathbf{C}$, we call the set consisting of all points

$$\alpha + t_1\omega_1 + t_2\omega_2 , \qquad 0 \leqslant t_i \leqslant 1$$

a **fundamental parallelogram** for the lattice (with respect to the given basis). We could also take the values $0 \leqslant t_i < 1$ to define a fundamental parallelogram, the advantage then being that in this case we get unique representatives for elements of $\mathbf{C}/L$ in $\mathbf{C}$.

Theorem 1.1. *Let P be a fundamental parallelogram for L, and assume that the elliptic function f has no poles on its boundary ∂P. Then the sum of the residues of f in P is* 0.

Proof. We have

$$2\pi i \sum \operatorname{Res} f = \int_{\partial P} f(z)\, dz = 0 ,$$

this last equality being valid because of the periodicity, so the integrals on opposite sides cancel each other.

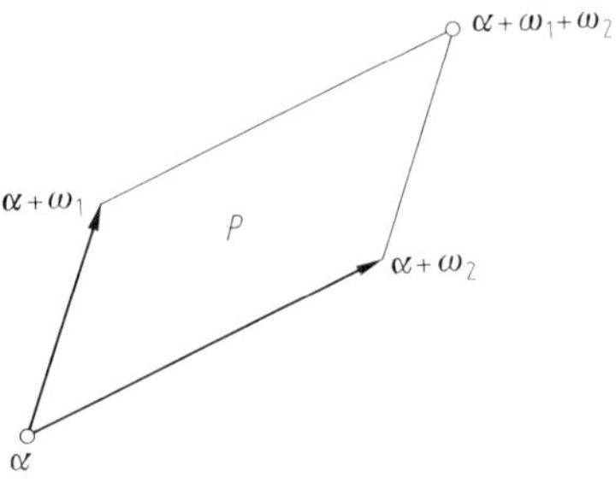

Fig. 2

An elliptic function can be viewed as a meromorphic function on the torus $\mathbf{C}/L$, and the above theorem can be interpreted as saying that the sum of the residues on the torus is equal to 0. Hence:

Corollary. *An elliptic function has at least two poles (counting multiplicities) on the torus.*

Theorem 1.2. *Let P be a fundamental parallelogram, and assume that the elliptic function f has no zero or pole on its boundary. Let $\{a_i\}$ be the singular points (zeros* *and poles) of f inside P, and let f have order m_i at a_i. Then*

$$\sum m_i = 0.$$

Proof. Observe that f elliptic implies that f' and f'/f are elliptic. We then obtain

$$0 = \int_{\partial P} f'/f(z)\,dz = 2\pi\sqrt{-1}\sum \text{Residues} = 2\pi\sqrt{-1}\sum m_i\,,$$

thus proving our assertion.

Again, we can formulate Theorem 1.2 by saying that the sum of the orders of the singular points of f on the torus is equal to 0.

Theorem 1.3. *Hypotheses being as in Theorem 1.2, we have*

$$\sum m_i a_i \equiv 0 \pmod{L}\,.$$

Proof. This time, we take the integral

$$\int_{\partial P} z\frac{f'(z)}{f(z)}\,dz = 2\pi\sqrt{-1}\sum m_i a_i,$$

because

$$\operatorname{res}_{a_i} z\frac{f'(z)}{f(z)} = m_i a_i\,.$$

On the other hand we compute the integral over the boundary of the parallelogram by taking it for two opposite sides at a time. One pair of such integrals is equal to

$$\int_{\alpha}^{\alpha+\omega_1} z\frac{f'(z)}{f(z)}\,dz - \int_{\alpha+\omega_2}^{\alpha+\omega_1+\omega_2} z\frac{f'(z)}{f(z)}\,dz\,.$$

We change variables in the second integral, letting $u = z - \omega_2$. Both integrals are then taken from α to $\alpha + \omega_1$, and after a cancellation, we get the value

$$-\omega_2\int_{\alpha}^{\alpha+\omega_1} \frac{f'(u)}{f(u)}\,du = 2\pi\sqrt{-1}\,k\omega_2,$$

for some integer k. The integral over the opposite pair of sides is done in the same way, and our theorem is proved.

§ 2. The Weierstrass Function

We now prove the existence of elliptic functions by writing some analytic expression, namely the Weierstrass function

$$\wp(z) = \frac{1}{z^2} + \sum_{\omega \in L'} \left[\frac{1}{(z-\omega)^2} - \frac{1}{\omega^2} \right],$$

where the sum is taken over the set of all non-zero periods, denoted by L'. We have to show that this series converges uniformly on compact sets not including the lattice points. For bounded z, staying away from the lattice points, the expression in the brackets has the order of magnitude of $1/|\omega|^3$. Hence it suffices to prove:

Lemma. *If* $\lambda > 2$, *then* $\sum_{\omega \in L'} \frac{1}{|\omega|^\lambda}$ *converges.*

Proof. The partial sum for $|\omega| \leqslant N$ can be decomposed into a sum for ω in the annulus at n, i.e. $n - 1 \leqslant |\omega| \leqslant n$, and then a sum for $1 \leqslant n \leqslant N$. In each annulus the number of lattice points has the order of magnitude n. Hence

$$\sum_{|\omega| \leqslant N} \frac{1}{|\omega|^\lambda} \ll \sum_1^\infty \frac{n}{n^\lambda} \ll \sum_1^\infty \frac{1}{n^{\lambda - 1}}$$

which converges for $\lambda > 2$.

The series expression for $\wp$ shows that it is meromorphic, with a double pole at each lattice point, and no other pole. It is also clear that $\wp$ is even, i.e.

$$\wp(z) = \wp(-z)$$

(summing over the lattice points is the same as summing over their negatives). We get $\wp'$ by differentiating term by term,

$$\wp'(z) = -2 \sum_{\omega \in L} \frac{1}{(z-\omega)^3},$$

the sum being taken for all $\omega \in L$. Note that $\wp'$ is clearly periodic, and is odd, i.e.

$$\wp'(-z) = -\wp'(z).$$

From its periodicity, we conclude that there is a constant C such that

$$\wp(z + \omega_1) = \wp(z) + C.$$

Let $z = -\omega_1/2$ (not a pole of $\wp$). We get

$$\wp\left(\frac{\omega_1}{2}\right) = \wp\left(-\frac{\omega_1}{2}\right) + C,$$

and since $\wp$ is even, it follows that $C = 0$. Hence $\wp$ is itself periodic, something which we could not see immediately from its series expansion.

It is clear that the set of all elliptic functions (with respect to a given lattice L) forms a field, whose constant field is the complex numbers.

Theorem 2.1. *The field of elliptic functions (with respect to L) is generated by $\wp$ and $\wp'$.*

Proof. If f is elliptic, we can write f as a sum of an even and an odd elliptic function as usual, namely

$$f(z) = \frac{f(z) + f(-z)}{2} + \frac{f(z) - f(-z)}{2}.$$

If f is odd, then the product $f\wp'$ is even, so it will suffice to prove that $\mathbf{C}(\wp)$ is the field of even elliptic functions, i.e. if f is even, then f is a rational function of $\wp$.

Suppose that f is even and has a zero of order m at some point u. Then clearly f also has a zero of the same order at $-u$ because

$$f^{(k)}(u) = (-1)^k f^{(k)}(-u).$$

Similarly for poles.

If $u \equiv -u \pmod{L}$, then the above assertion holds in the strong sense, namely f has a zero (or pole) of even order at u.

Proof. First note that $u \equiv -u \pmod{L}$ is equivalent to

$$2u \equiv 0 \pmod{L}.$$

On the torus, there are exactly four points with this property, represented by

$$0, \frac{\omega_1}{2}, \frac{\omega_2}{2}, \frac{\omega_1 + \omega_2}{2}$$

in a period parallelogram. If f is even, then f' is odd, i.e.

$$f'(u) = -f'(-u).$$

Since $u \equiv -u \pmod{L}$ and f' is periodic, it follows that $f'(u) = 0$, so that f has a zero of order at least 2 at u. If $u \not\equiv 0 \pmod{L}$, then the above argument shows that the function

$$g(z) = \wp(z) - \wp(u)$$

has a zero of order at least 2 (hence exactly 2 by Theorem 1.2 and the fact that $\wp$ has only one pole of order 2 on the torus). Then f/g is even, elliptic, holomorphic at u. If $f(u)/g(u) \neq 0$ then $\text{ord}_u f = 2$. If $f(u)/g(u) = 0$ then f/g again has a zero of order at least 2 at u and we can repeat the argument. If $u \equiv 0 \pmod{L}$ we use $g = 1/\wp$ and argue similarly, thus proving that f has a zero of even order at u. For the case of poles, we consider $1/f$ instead of f.

Now let u_i $(i = 1, \ldots, r)$ be a family of points containing one representative from each class $(u, -u) \pmod{L}$ where f has a zero or pole, other than the class of L itself. Let

$$m_i = \text{ord}_{u_i} f \quad \text{if } 2u_i \not\equiv 0 \pmod{L},$$
$$m_i = \tfrac{1}{2}\,\text{ord}_{u_i} f \quad \text{if } 2u_i \equiv 0 \pmod{L}.$$

Our previous remarks show that for $a \in \mathbf{C}$, $a \not\equiv 0 \pmod{L}$, the function $\wp(z) - \wp(a)$ has a zero of order 2 at a if and only if $2a \equiv 0 \pmod{L}$, and has distinct zeros of order 1 at a and $-a$ otherwise. Hence for all $z \not\equiv 0 \pmod{L}$ the function

$$\prod_{i=1}^{r} [\wp(z) - \wp(u_i)]^{m_i}$$

has the same order at z as f. This is also true at the origin because of Theorem 1.2 applied to f and the above product. The quotient of the above product by f is then an elliptic function without zero or pole, hence a constant, thereby proving Theorem 2.1.

Next, we obtain the power series development of $\wp$ and $\wp'$ at the origin, from which we shall get the algebraic relation holding between these two functions. We do this by brute force.

$$\begin{aligned}
\wp(z) &= \frac{1}{z^2} + \sum_{\omega \in L'} \left[\frac{1}{\omega^2} \left(1 + \frac{z}{\omega} + \left(\frac{z}{\omega}\right)^2 + \cdots \right)^2 - \frac{1}{\omega^2} \right] \\
&= \frac{1}{z^2} + \sum_{\omega \in L'} \sum_{m=1}^{\infty} (m+1) \left(\frac{z}{\omega}\right)^m \frac{1}{\omega^2} \\
&= \frac{1}{z^2} + \sum_{m=1}^{\infty} c_m z^m
\end{aligned}$$

where

$$c_m = \sum_{\omega \neq 0} \frac{m+1}{\omega^{m+2}}.$$

Note that $c_m = 0$ if m is odd.

Using the notation

$$s_m(L) = s_m = \sum_{\omega \neq 0} \frac{1}{\omega^m}$$

we get the expansion

$$\wp(z) = \frac{1}{z^2} + \sum_{n=1}^{\infty} (2n+1)s_{2n+2}(L)z^{2n},$$

from which we write down the first few terms explicitly:

$$\wp(z) = \frac{1}{z^2} + 3s_4 z^2 + 5s_6 z^4 + \cdots$$

and differentiating term by term,

$$\wp'(z) = \frac{-2}{z^3} + 6s_4 z + 20s_6 z^3 + \cdots$$

Theorem 2.2. *Let* $g_2 = g_2(L) = 60s_4$ *and* $g_3 = g_3(L) = 140s_6$. *Then*

$$\wp'^2 = 4\wp^3 - g_2\wp - g_3 .$$

Proof. We expand out the function

$$\varphi(z) = \wp'(z)^2 - 4\wp(z)^3 + g_2\wp(z) + g_3$$

at the origin, paying attention only to the polar term and the constant term. This is easily done, and one sees that there is enough cancellation so that these terms are 0, in other words, $\varphi(z)$ is an elliptic function without poles, and with a zero at the origin. Hence φ is identically zero, thereby proving our theorem.

The preceding theorem shows that the points $(\wp(z), \wp'(z))$ lie on the curve defined by the equation

$$y^2 = 4x^3 - g_2x - g_3 .$$

The cubic polynomial on the right-hand side has a discriminant given by

$$\Delta(\omega_1, \omega_2) = \Delta = g_2^3 - 27g_3^2 .$$

We shall see in a moment that this discriminant does not vanish.

Let

$$e_i = \wp\left(\frac{\omega_i}{2}\right), \qquad i = 1, 2, 3,$$

where $L = [\omega_1, \omega_2]$ and $\omega_3 = \omega_1 + \omega_2$. Then the function

$$h(z) = \wp(z) - e_i$$

has a zero at $\omega_i/2$, which is of even order so that $\wp'(\omega_i/2) = 0$ for $i = 1, 2, 3$, by previous remarks. Comparing zeros and poles, we conclude that

$$\wp'^2(z) = 4(\wp(z) - e_1)(\wp(z) - e_2)(\wp(z) - e_3).$$

Thus e_1, e_2, e_3 are the roots of $4x^3 - g_2x - g_3$. Furthermore, $\wp$ takes on the value e_i with multiplicity 2 and has only one pole of order 2 mod L, so that $e_i \neq e_j$ for $i \neq j$. This means that the three roots of the cubic polynomial are distinct, and therefore

$$\Delta = g_2^3 - 27g_3^2 \neq 0.$$

§ 3. The Addition Theorem

Given complex numbers g_2, g_3 such that $g_2^3 - 27g_3^2 \neq 0$, one can ask whether there exists a lattice for which these are the invariants associated to the lattice as in the preceding section. The answer is yes, but here we consider the case when g_2, g_3 are given as in the preceding section, i.e. $g_2 = 60s_4$ and $g_3 = 140s_6$.

We have seen that the map

$$z \mapsto (1, \wp(z), \wp'(z))$$

parametrizes points on the cubic curve A defined by the equation

$$y^2 = 4x^3 - g_2x - g_3.$$

This is an affine equation, and we put in the coordinate 1 to indicate that we also view the points as embedded in projective space. Then the mapping is actually defined on the torus $\mathbf{C}/L$, and the lattice points, i.e. 0 on the torus, are precisely the points going to infinity on the curve. Let $A_{\mathbf{C}}$ denote the complex points on the curve. We in fact get a bijection

$$\mathbf{C}/L - \{0\} \to A_{\mathbf{C}} - \{\infty\}.$$

This is easily seen: For any complex number α, $\wp(z) - \alpha$ has at most two zeros, and at least one zero, so that already under $\wp$ we cover each complex number α. It is then verified at once that using $\wp'$ separates the points of $\mathbf{C}/L$ lying above α, thus giving us the bijection. If you know the terminology of algebraic geometry, then you know that the curve defined by the above equation is non-singular, and that our mapping is actually a complex analytic isomorphism between $\mathbf{C}/L$ and $A_{\mathbf{C}}$.

Furthermore, $\mathbf{C}/L$ has a natural group structure, and we now want to see what it looks like when transported to A. We shall see that it is algebraic. In other words, if

$$P_1 = (x_1, y_1), \qquad P_2 = (x_2, y_2), \qquad P_3 = (x_3, y_3)$$

and

$$P_3 = P_1 + P_2 ,$$

then we shall express x_3, y_3 as rational functions of (x_1, y_1) and (x_2, y_2). We shall see that P_3 is obtained by taking the line through P_1, P_2, intersecting it with the curve, and reflecting the point of intersection through the x-axis, as shown on Fig. 3.

Select $u_1, u_2 \in \mathbf{C}$ and $\notin L$, and assume $u_1 \not\equiv u_2 \pmod L$. Let a, b be complex numbers such that

$$\wp'(u_1) = a\wp(u_1) + b$$

$$\wp'(u_2) = a\wp(u_2) + b ,$$

in other words $y = ax + b$ is the line through $(\wp(u_1), \wp'(u_1))$ and $(\wp(u_2), \wp'(u_2))$. Then

$$\wp'(z) - (a\wp(z) + b)$$

has a pole of order 3 at 0, whence it has three zeros, counting multiplicities, and two of these are at u_1 and u_2. If, say, u_1 had multiplicity 2, then by Theorem 1.3 we would have

$$2u_1 + u_2 \equiv 0 \pmod L) .$$

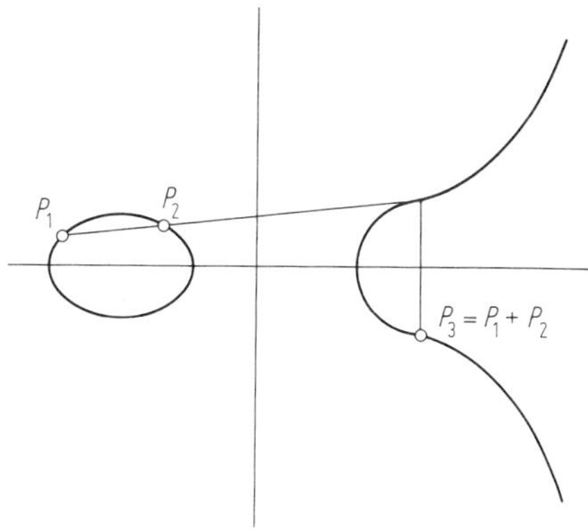

Fig. 3

If we fix u_1, this can hold for only one value of u_2. Let us assume that we do not deal with this value. Then both u_1, u_2 have multiplicity 1, and the third zero lies at

$$u_3 \equiv -(u_1 + u_2) \pmod{L}$$

again by Theorem 1.3. So we also get

$$\wp'(u_3) = a\wp(u_3) + b\,.$$

The equation

$$4x^3 - g_2x - g_3 - (ax + b)^2 = 0$$

has three roots, counting multiplicities. They are $\wp(u_1)$, $\wp(u_2)$, $\wp(u_3)$, and the left-hand side factors as

$$4(x - \wp(u_1))(x - \wp(u_2))(x - \wp(u_3))\,.$$

Comparing the coefficient of x^2 yields

$$\wp(u_1) + \wp(u_2) + \wp(u_3) = \frac{a^2}{4}\,.$$

But from our original equations for a and b, we have

$$a(\wp(u_1) - \wp(u_2)) = \wp'(u_1) - \wp'(u_2)\,.$$

Therefore from

$$\wp(u_3) = \wp(-(u_1 + u_2)) = \wp(u_1 + u_2)$$

we get

$$\boxed{\wp(u_1 + u_2) = -\wp(u_1) - \wp(u_2) + \frac{1}{4}\left(\frac{\wp'(u_1) - \wp'(u_2)}{\wp(u_1) - \wp(u_2)}\right)^2}$$

or in algebraic terms,

$$x_3 = -x_1 - x_2 + \frac{1}{4}\left(\frac{y_1 - y_2}{x_1 - x_2}\right)^2.$$

Fixing u_1, the above formula is true for all but a finite number of $u_2 \not\equiv u_1 \pmod{L}$, whence for all $u_2 \not\equiv u_1 \pmod{L}$ by analytic continuation.

For $u_1 \equiv u_2 \pmod{L}$ we take the limit as $u_1 \to u_2$ and get

$$\wp(2u) = -2\wp(u) + \frac{1}{4}\left(\frac{\wp''(u)}{\wp'(u)}\right)^2 .$$

These give us the desired algebraic addition formulas. Note that the formulas involve only g_2, g_3 as coefficients in the rational functions.

§ 4. Endomorphisms, Automorphisms, and Isomorphisms

Theorem 4.1. *Let L, M be two lattices in* $\mathbf{C}$ *and let*

$$\lambda : \mathbf{C}/L \to \mathbf{C}/M$$

be a complex analytic homomorphism. Then there exists a complex number α *such that the following diagram is commutative.*

$$\begin{array}{ccc} \mathbf{C} & \xrightarrow{\alpha} & \mathbf{C} \\ \downarrow & & \downarrow \\ \mathbf{C}/L & \xrightarrow[\lambda]{} & \mathbf{C}/M \end{array}$$

The top map is multiplication by α*, and the vertical maps are the canonical homomorphisms.*

Proof. Locally near 0, λ can be expressed by a power series.

$$\lambda(z) = a_0 + a_1 z + a_2 z^2 + \cdots,$$

and since a complex number near 0 represents uniquely its class mod L, it follows from the formula

$$\lambda(z + z') \equiv \lambda(z) + \lambda(z') \pmod{M}$$

that the congruence can actually be replaced by an equality. Hence we must have

$$\lambda(z) = a_1 z ,$$

for z near 0. But z/n for arbitrary z and large n is near 0, and from this one concludes that for any z we must have

$$\lambda(z) \equiv a_1 z \pmod{M} .$$

This proves our theorem.

We see that λ is represented by a multiplication α, and that

$$\alpha L \subset M .$$

Conversely, given a complex number α and lattices L, M such that $\alpha L \subset M$, multiplication by α induces a complex analytic homomorphism of $\mathbf{C}/L$ into $\mathbf{C}/M$.

Two complex toruses $\mathbf{C}/L$ and $\mathbf{C}/M$ are isomorphic if and only if there exists a complex number α such that $\alpha L = M$. We shall say that two lattices L, M are **linearly equivalent** if this condition is satisfied.

By an **elliptic curve**, or **abelian curve** A, one means a complete non-singular curve of genus 1, and a special point O taken as origin. The Riemann–Roch theorem defines a group law on the group of divisor classes of A. Actually, if P, P' are points on A, then there exists a unique point P'' such that

$$(P) + (P') \sim (P'') + (O) ,$$

where $\sim$ means linear equivalence, i.e. the left-hand side minus the right-hand side is the divisor of a rational function on the curve. The group law on A is then

$$P + P' = P'' .$$

In characteristic $\neq 2$ or 3, using the Riemann–Roch theorem, one finds that the curve can be defined by a Weierstrass equation

$$y^2 = 4x^3 - g_2 x - g_3 ,$$

with g_2, g_3 in the ground field over which the curve is defined. Conversely, any homogeneous non-singular cubic equation has genus 1 and defines an abelian curve in the projective plane, once the origin has been selected. These facts depend on elementary considerations of curves.

A curve defined by equations in projective space is said to be **defined over a field** k if the coefficients of these equations lie in k. For the Weierstrass equation, this means $g_2, g_3 \in k$.

For our purposes, if the reader is willing to exclude certain special cases, it will always suffice to visualize an elliptic curve as a curve defined by the above equation, with the addition law given by the rational formulas obtained from the addition theorem of the $\wp$ function. The origin is then the point at infinity. If A is defined over k, we denote by A_k the set of points (x, y) on the curve with $x, y \in k$, together with infinity, and call it the group of k-**rational points** on the curve. It is a group because the addition is rational, with coefficients in k.

If A, B are elliptic curves, one calls a **homomorphism** of A into B a group homomorphism whose graph is algebraic in the product space. If $\lambda\colon A \to B$ is such a homomorphism, and the curves are defined over the complex numbers, then λ

induces a complex analytic homomorphism also denoted by λ,

$$\lambda: A_{\mathbf{C}} \to B_{\mathbf{C}},$$

viewing the groups of complex points on A and B as complex analytic groups. Suppose that the curves are obtained from lattices L and M in $\mathbf{C}$ respectively, i.e. we have maps

$$\varphi: \mathbf{C}/L \to A_{\mathbf{C}} \quad \text{and} \quad \psi: \mathbf{C}/M \to B_{\mathbf{C}}$$

which are analytic isomorphisms. As we saw above, our homomorphism λ is then induced by a multiplication by a complex number.

Conversely, it can be shown that any complex analytic homomorphism

$$\gamma: \mathbf{C}/L \to \mathbf{C}/M$$

induces an algebraic one, i.e. there exists an algebraic homomorphism λ which makes the following diagram commutative.

$$\begin{array}{ccc} \mathbf{C}/L & \xrightarrow{\gamma} & \mathbf{C}/M \\ \varphi \downarrow & & \downarrow \psi \\ A_{\mathbf{C}} & \xrightarrow[\lambda]{} & B_{\mathbf{C}} \end{array}$$

We shall make a table of the effect of an isomorphism on the coefficients of the equations for elliptic curves, and their coordinates.

Let us agree that if A is an elliptic curve parametrized by the Weierstrass functions, for the rest of this section,

$$\varphi_A: \mathbf{C}/L \to A_{\mathbf{C}}$$

is the map such that

$$\varphi_A(z) = (1, \wp(z), \wp'(z)).$$

The $\wp$ function depends on L, and we shall denote it by

$$\wp(z, L).$$

Similarly for $\wp'(z, L)$. These satisfy the homogeneity property

$$\wp(cz, cL) = c^{-2}\wp(z, L) \quad \text{and} \quad \wp'(cz, cL) = c^{-3}\wp'(z, L)$$

for any $c \in \mathbf{C}$, $c \neq 0$.

Suppose that we are given two elliptic curves with parametrizations

$$\varphi_A : \mathbf{C}/L \to A_{\mathbf{C}} \quad \text{and} \quad \varphi_B : \mathbf{C}/M \to B_{\mathbf{C}},$$

and suppose that

$$M = cL,$$

so that the curves are isomorphic, with an isomorphism

$$\lambda : A \to B$$

induced by the multiplication by c. Then the coefficients g_2, g_3 of these curves satisfy the transformation

$$g_2(cL) = c^{-4} g_2(L)$$
$$g_3(cL) = c^{-6} g_3(L).$$

We let x_A and x_B denote the x-coordinate in the Weierstrass equation satisfied by the curves, respectively. Thus in general,

$$x(\varphi(z)) = \wp(z),$$

and similarly

$$y(\varphi(z)) = \wp'(z).$$

If P is a point on A, then the homogeneity properties of the Weierstrass functions can then be expressed purely algebraically by the formulas

$$x_B(\lambda(P)) = c^{-2} x_A(P) \quad \text{and} \quad y_B(\lambda(P)) = c^{-3} y_A(P).$$

If $L = M$, we get all endomorphisms (complex analytic) of $\mathbf{C}/L$ by those complex α such that $\alpha L \subset L$. Those endomorphisms induced by ordinary integers are called **trivial**. In general, suppose that $L = [\omega_1, \omega_2]$ and $\alpha L \subset L$. Then there exist integers a, b, c, d such that

$$\alpha\omega_1 = a\omega_1 + b\omega_2,$$
$$\alpha\omega_2 = c\omega_1 + d\omega_2.$$

Therefore α is a root of the polynomial equation

$$\begin{vmatrix} x - a & -b \\ -c & x - d \end{vmatrix} = 0,$$

whence we see that α is quadratic over $\mathbf{Q}$, and is in fact integral over $\mathbf{Z}$. Dividing $\alpha\omega_2$ by ω_2, we see that

$$\alpha = c\tau + d,$$

where $\tau = \omega_1/\omega_2$. Since ω_1, ω_2 span a lattice, their ratio cannot be real. If α is not an integer, then $c \neq 0$, and consequently

$$\mathbf{Q}(\tau) = \mathbf{Q}(\alpha).$$

Furthermore, α is not real, i.e. α is imaginary quadratic.

The ring R of elements $\alpha \in \mathbf{Q}(\tau)$ such that $\alpha L \subset L$ is a subring of the quadratic field $k = \mathbf{Q}(\tau)$, and is in fact a subring of the ring of all algebraic integers $\mathfrak{o}_k$ in k. The units in R represent the automorphisms of $\mathbf{C}/L$. It is well known and very easy to prove that in imaginary quadratic field, the only units of R are roots of unity, and a quadratic field contains roots of unity other than ± 1 if and only if

$$k = \mathbf{Q}(\sqrt{-1}) \quad \text{or} \quad k = \mathbf{Q}(\sqrt{-3}).$$

If R contains $i = \sqrt{-1}$, then $R = \mathbf{Z}[i]$ is the ring of all algebraic integers in k, which must be $\mathbf{Q}(i)$. If R contains a cube root of unity ρ, then $R = \mathbf{Z}[\rho]$ is the ring of all algebraic integers in k, which must be $\mathbf{Q}(\sqrt{-3})$. The units in this ring are the 6-th roots of unity, generated by $-\rho$.

§ 5. Points of Finite Order

Let A be an elliptic curve defined over a field k. For each positive integer N we denote by A_N the kernel of the map

$$t \mapsto Nt, \quad t \in A,$$

i.e. it is the subgroup of points of order N. If A is defined over the complex numbers, then it is immediately clear from the representation $A_{\mathbf{C}} \approx \mathbf{C}/L$ that

$$A_N \approx \mathbf{Z}/N\mathbf{Z} \times \mathbf{Z}/N\mathbf{Z}.$$

The inverse image of these points in $\mathbf{C}$ occur as the points of the lattice $\frac{1}{N}L$, and their inverse image in $\mathbf{C}/L$ is therefore the subgroup

$$\frac{1}{N}L/L \subset \mathbf{C}/L.$$

Let

$$\varphi: \mathbf{C} \to A_{\mathbf{C}}$$

be an analytic representation of $A_{\mathbf{C}}$ as $\mathbf{C}/L$, and let $L = [\omega_1, \omega_2]$. If we let

$$t_1 = \varphi\left(\frac{\omega_1}{N}\right) \quad \text{and} \quad t_2 = \varphi\left(\frac{\omega_2}{N}\right),$$

then $\{t_1, t_2\}$ form a basis for A_N over $\mathbf{Z}/N\mathbf{Z}$, i.e. A_N is the direct sum of the cyclic groups of order N generated by t_1 and t_2, respectively.

If the elliptic curve is defined over a field of characteristic zero, say k finitely generated over the rationals, then we can embed k in $\mathbf{C}$ and apply the preceding result.

Let A be an elliptic curve defined over a field k and let K be an extension of k. Let σ be an isomorphism of K, not necessarily identity on k. One defines A^σ to be the curve obtained by applying σ to the coefficients of the equation defining A. For instance, if A is defined by

$$y^2 = 4x^3 - g_2 x - g_3,$$

then A^σ is defined by

$$y^2 = 4x^3 - g_2^\sigma x - g_3^\sigma.$$

If P, Q are points of A in K, then we have the formula

$$(P + Q)^\sigma = P^\sigma + Q^\sigma.$$

The sum on the left refers to addition on A, and the sum on the right refers to addition on A^σ. This is obvious because the algebraic addition formula is given by rational functions in the coordinates, with coefficients in k. Of course, if $P = (x, y)$, then $P^\sigma = (x^\sigma, y^\sigma)$ is obtained by applying σ to the coordinates.

In particular, suppose that P is a point of finite order, so that $NP = O$. Since O is rational over k, we see that for any isomorphism σ of K over k we have $NP^\sigma = O$ also, whence P^σ is also a point of order N. Since the number of points of order N is finite, it follows in particular that the points of A_N are algebraic over k (i.e. their coordinates are algebraic over k).

If $P = (x, y)$, we let $k(P) = k(x, y)$ be the extension of k obtained by adjoining the coordinates of P. Similarly, we let

$$k(A_N)$$

be the compositum of all fields $k(P)$ for $P \in A_N$. Of course, we view all points of finite order as having coordinates in a fixed algebraic closure of k, which we denote by ${}^a k$ or k_a.

The above remarks show that the Galois group $\mathrm{Gal}(k_a/k)$ operates as a permutation group of A_N. Consequently $k(A_N)$ is a normal extension of k, and is Galois if N is not divisible by the characteristic of k. We call $k(A_N)$ the **field of N-division points** of A over k.

Furthermore, if σ is an automorphism of $k(A_N)$ over k, and if we let $\{t_1, t_2\}$ be a basis of A_N over $\mathbf{Z}/N\mathbf{Z}$, then σ can be represented by a matrix

$$\begin{pmatrix} a & b \\ c & d \end{pmatrix}$$

such that

$$\begin{pmatrix} \sigma t_1 \\ \sigma t_2 \end{pmatrix} = \begin{pmatrix} at_1 + bt_2 \\ ct_1 + dt_2 \end{pmatrix} = \begin{pmatrix} a & b \\ c & d \end{pmatrix} \begin{pmatrix} t_1 \\ t_2 \end{pmatrix}.$$

Thus we get an injective homomorphism

$$\mathrm{Gal}(k(A_N)/k) \to GL_2(\mathbf{Z}/N\mathbf{Z}) .$$

§ 6. The Sigma and Zeta Function

Both in number theory and analysis one factorizes elements into prime powers. In analysis, this means that a function gets factored into an infinite product corresponding to its zeros and poles. Taking the values at special points, such an analytic expression reflects itself into special properties of the values, for which it becomes possible to determine the prime factorization in number fields.

In this section, we are concerned with the analytic expressions.

Our first task is to give a universal gadget allowing us to factorize an elliptic function, with a numerator and denominator which are entire functions, and are as periodic as possible.

One defines a **theta function** (on $\mathbf{C}$) with respect to a lattice L, to be an entire function θ satisfying the condition

$$\theta(z + u) = \theta(z)\, e^{2\pi i[l(z,u) + c(u)]}, \qquad z \in \mathbf{C},\ u \in L,$$

where l is $\mathbf{C}$-linear in z, $\mathbf{R}$-linear in u, and $c(u)$ is some function depending only on u. We shall construct a theta function.

We write down the **Weierstrass sigma function**, which has zeros of order 1 at all lattice points, by the Weierstrass product

$$\sigma(z) = z \prod_{\omega \in L'} \left(1 - \frac{z}{\omega}\right) e^{z/\omega + \frac{1}{2}(z/\omega)^2} .$$

Here L' means the lattice from which 0 is deleted, i.e. we are taking the product over the non-zero periods. We note that σ also depends on L, and so we write $\sigma(z, L)$, which is homogeneous of degree 1, namely

$$\boxed{\sigma(\lambda z, \lambda L) = \lambda\sigma(z, L)} \qquad \lambda \in \mathbf{C}.$$

Taking the logarithmic derivative formally yields the **Weierstrass zeta function**

$$\zeta(z, L) = \zeta(z) = \frac{\sigma'(z)}{\sigma(z)} = \frac{1}{z} + \sum_{\omega \in L'} \left[\frac{1}{z - \omega} + \frac{1}{\omega} + \frac{z}{\omega^2}\right].$$

It is clear that the sum on the right converges absolutely and uniformly for z in a compact set not containing any lattice point, and hence integrating and exponentiating shows that the infinite product for $\sigma(z)$ also converges absolutely and uniformly in such a region. Differentiating $\zeta(z)$ term by term shows that

$$\zeta'(z) = -\wp(z) = -\frac{1}{z^2} - \sum_{\omega \in L'} \left[\frac{1}{(z - \omega)^2} - \frac{1}{\omega^2}\right].$$

Also from the product and sum expressions, we see at once that *both σ and ζ are odd functions*, i.e.

$$\sigma(-z) = -\sigma(z) \quad \text{and} \quad \zeta(-z) = -\zeta(z).$$

The series defining $\zeta(z, L)$ shows that it is homogenous of degree -1, that is

$$\boxed{\zeta(\lambda z, \lambda L) = \frac{1}{\lambda}\zeta(z, L).}$$

Differentiating the function $\zeta(z + \omega) - \zeta(z)$ for any $\omega \in L$ yields 0 because the $\wp$-function is periodic. Hence there is a constant $\eta(\omega)$ (sometimes written η_ω) such that

$$\zeta(z + \omega) = \zeta(z) + \eta(\omega).$$

It is clear that $\eta(\omega)$ is $\mathbf{Z}$-linear in ω. If $L = [\omega_1, \omega_2]$, then one uses the notation

$$\eta(\omega_1) = \eta_1 \quad \text{and} \quad \eta(\omega_2) = \eta_2.$$

As with ζ, the form $\eta(\omega)$ satisfies the homogeneity relation as one verifies directly from the similar relation for ζ. Observe that the lattice should strictly be in the notation, so that in full, the relations should read

$$\boxed{\begin{aligned} &\zeta(z + \omega, L) = \zeta(z, L) + \eta(\omega, L) \\ &\eta(\lambda\omega, \lambda L) = \frac{1}{\lambda}\eta(\omega, L). \end{aligned}}$$

Theorem 6.1. *The function σ is a theta function, and in fact*

$$\frac{\sigma(z+\omega)}{\sigma(z)} = \psi(\omega)\, e^{\eta(\omega)(z+\omega/2)}$$

where

$$\psi(\omega) = 1 \qquad \text{if } \omega/2 \in L$$

$$\psi(\omega) = -1 \quad \text{if } \omega/2 \notin L\,.$$

Proof. We have

$$\frac{d}{dz} \log \frac{\sigma(z+\omega)}{\sigma(z)} = \eta(\omega)\,.$$

Hence

$$\log \frac{\sigma(z+\omega)}{\sigma(z)} = \eta(\omega)z + c(\omega)\,,$$

whence exponentiating yields

$$\sigma(z+\omega) = \sigma(z)\, e^{\eta(\omega)z + c(\omega)}\,,$$

which shows that σ is a theta function. We write the quotient as in the statement of the theorem, thereby defining $\psi(\omega)$, and it is then easy to determine $\psi(\omega)$ as follows.

Suppose that $\omega/2$ is not a period. Set $z = -\omega/2$ in the above relation. We see at once that $\psi(\omega) = -1$ because σ is odd. On the other hand, consider

$$\frac{\sigma(z+2\omega)}{\sigma(z)} = \frac{\sigma(z+2\omega)}{\sigma(z+\omega)} \frac{\sigma(z+\omega)}{\sigma(z)}\,.$$

Using the functional equation twice and comparing the two sides, we see that $\psi(2\omega) = \psi(\omega)^2$. In particular, if $\omega/2 \in L$, then

$$\psi(\omega) = \psi(\omega/2)^2\,.$$

Dividing by 2 until we get some element of the lattice which is not equal to twice a period, we conclude at once that $\psi(\omega) = (-1)^{2n} = 1$.

The numbers η_1 and η_2 are called **basic quasi periods of** ζ.

Legendre Relation. *We have*

$$\eta_2\omega_1 - \eta_1\omega_2 = 2\pi i\,.$$

Proof. We integrate around a fundamental parallelogram P, just as we did for the $\wp$-function:

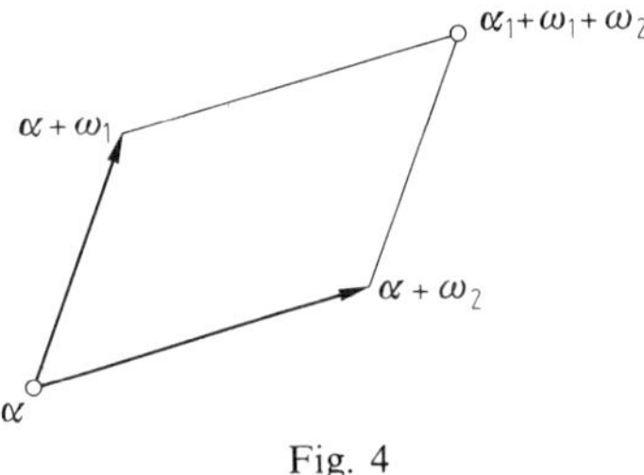

Fig. 4

The integral is equal to

$$\int_{\partial P} \zeta(z)\,dz = 2\pi i \sum \text{residues of } \zeta$$

$$= 2\pi i$$

because ζ has residue 1 at 0 and no other pole in a fundamental parallelogram containing 0. On the other hand, using the quasi periodicity, the integrals over opposite sides combine to give

$$\eta_2\omega_1 - \eta_1\omega_2\,,$$

as desired.

Next, we show how the sigma function can be used to factorize elliptic functions. We know that the sum of the zeros and poles of an elliptic function must be congruent to zero modulo the lattice. Selecting suitable representatives of these zeros and poles, we can always make the sum equal to 0.

For any $a \in \mathbf{C}$ we have

$$\frac{\sigma(z+a+\omega)}{\sigma(z+a)} = \psi(\omega)\, e^{\eta(\omega)(z+\omega/2)}\, e^{\eta(\omega)a}\,.$$

Observe how the term $\eta(\omega)a$ occurs linearly in the exponent. It follows that if $\{a_i\}$, $\{b_i\}$ $(i = 1,\ldots,n)$ are families of complex numbers such that

$$\sum a_i = \sum b_i\,,$$

then the function

$$\frac{\prod \sigma(z-a_i)}{\prod \sigma(z-b_i)}$$

is periodic with respect to our lattice, and is therefore an elliptic function. Conversely, any elliptic function can be so factored into a numerator and denominator involving the sigma function. We write down explicitly the special case with the $\wp$-function.

Theorem 6.2. *For any $a \in \mathbf{C}$ not in L, we have*

$$\wp(z) - \wp(a) = -\frac{\sigma(z+a)\sigma(z-a)}{\sigma^2(z)\sigma^2(a)}.$$

Proof. The function $\wp(z) - \wp(a)$ has zeros at a and $-a$, and has a double pole at 0. Hence

$$\wp(z) - \wp(a) = C\frac{\sigma(z+a)\sigma(z-a)}{\sigma^2(z)}$$

for some constant C. Multiply by z^2 and let $z \to 0$. Then $\sigma^2(z)/z^2$ tends to 1 and $z^2\wp(z)$ tends to 1. Hence we get the value $C = -1/\sigma^2(a)$, thus proving our theorem.

§ 7. The Klein Form and the Siegel–Néron Function

The function $\omega \mapsto \eta(\omega)$ extends by $\mathbf{R}$-linearity to any complex number, so that if u_1, u_2 are real numbers, we put

$$u = u_1\omega_1 + u_2\omega_2, \qquad \eta(u) = u_1\eta_1 + u_2\eta_2.$$

We define the **Klein form**

$$\mathfrak{k}(u; \omega_1, \omega_2) = e^{-\eta(u)u/2}\sigma(u; \omega_1, \omega_2).$$

It is homogeneous of degree 1 in the sense that for all $\lambda \in \mathbf{C}^*$,

K 0. $$\mathfrak{k}(u_1, u_2; \lambda\omega_1, \lambda\omega_2) = \lambda\mathfrak{k}(u_1, u_2; \omega_1, \omega_2).$$

Furthermore, if $\alpha \in SL_2(\mathbf{Z})$ then

K 1. $$\mathfrak{k}\left((u_1, u_2); \alpha\begin{pmatrix}\omega_1\\ \omega_2\end{pmatrix}\right) = \mathfrak{k}\left((u_1, u_2)\alpha; \begin{pmatrix}\omega_1\\ \omega_2\end{pmatrix}\right).$$

When u_1, u_2 are rational numbers, the Klein form is that considered in [KL], and is modular of a certain level. In the present case with arbitrary real u_1, u_2, it is merely real analytic.

Next we have a property which shows that $\mathfrak{k}$ is nearly periodic.

K 2(i) $$\mathfrak{k}(u+\omega_1) = -e^{-2\pi i u_2/2}\mathfrak{k}(u)$$

K 2(ii) $$\mathfrak{k}(u+\omega_2) = -e^{2\pi i u_1/2}\mathfrak{k}(u)\,.$$

These two formulas are immediate by using the definitions and the Legendre relation.

The important consequence for us is that the function

$$u \mapsto \operatorname{Re}\log \mathfrak{k}(u) = \log|\mathfrak{k}(u)|$$

is periodic, so defined on $\mathbf{C}/L$, *in other words, defined on the elliptic curve.*

We now define the **Siegel function**

$$\begin{aligned} g(u) &= \Delta(\omega_1,\omega_2)^{1/12}\mathfrak{k}(u;\omega_1,\omega_2) \\ &= \Delta(\omega_1,\omega_2)^{1/12}\, e^{-\eta(u)u/2}\sigma(u;L)\,, \end{aligned}$$

and the local **Néron function**

$$\boxed{\begin{aligned} \lambda(u) &= -\operatorname{Re}\log g(u) = -\log|g(u)| \\ &= -\log|\sigma(u;\omega_1,\omega_2)| + \tfrac{1}{2}\operatorname{Re}\eta(u)u - \tfrac{1}{12}\log|\Delta(L)|\,. \end{aligned}}$$

(Cf. Néron [Ne 1], Chapter III, § 7, last page of the paper.) Because of minus signs, and to relate with the order function in the non-archimedean case, it is useful to define

$$v(t) = -\log|t|\,.$$

Then

$$\boxed{\lambda(u) = v(\mathfrak{k}(u)) + \tfrac{1}{12}v(\Delta) = v(g(u))\,.}$$

Theorem 7.1. *The Néron function satisfies the relation*

$$\lambda(u+w) + \lambda(u-w) = 2\lambda(u) + 2\lambda(w) + v(\wp(u)-\wp(w)) - \tfrac{1}{6}v(\Delta)\,,$$

for any complex numbers u, w *such that* u, w, $u \pm w$ *are not lattice points.*

Proof. This is immediate from the expression for the Weierstrass function in terms of the sigma function given in Theorem 6.2, and the linearity properties of the quadratic form

$$u \mapsto \tfrac{1}{2} \operatorname{Re} \eta(u)u\,,$$

which causes the extraneous terms to cancel.

The function λ can be characterized easily. Recall that an (analytic) local parameter at the origin O of the elliptic curve A is an analytic function in a neighborhood of O which has a zero of order 1 at O. Since λ depends only on the point u in $\mathbf{C}/L$, we shall also write λ as function of points on $A_{\mathbf{C}}$. Thus if

$$P = \exp u$$

we also write $P = P_u$ and $\lambda(u) = \lambda(P_u) = \lambda(P)$.

Theorem 7.2. *The Néron function λ is the unique function on $A_{\mathbf{C}} - \{O\}$, satisfying the following conditions:*

(i) *λ is continuous and bounded outside every neighborhood of O.*

(ii) *For one (and therefore every) local parameter z at O, the limit*

$$\lim_{P \to O} \lambda(P) + \log |z(P)|$$

exists.

(iii) *For all $P, Q \in A_{\mathbf{C}}$ such that $P, Q, P \pm Q \neq O$ on $A_{\mathbf{C}}$, we have*

$$\lambda(P+Q) + \lambda(P-Q) = 2\lambda(P) + 2\lambda(Q) + v(x(P) - x(Q)) - \tfrac{1}{6}v(\Delta).$$

Proof. Property (i) is obvious from the definition of λ and the periodicity (looking at values of u staying away from 0). Since the sigma function is an analytic local parameter at the origin, the expression for λ in terms of σ shows that (ii) is satisfied. Any two local parameters differ by an invertible analytic function at the origin, so the existence of a limit is independent of the parameter. Condition (iii) was proved in the preceding theorem.

As for uniqueness, let f be the difference of two functions satisfying the three conditions. Then f can be defined at 0 by continuity, and is thus continuous everywhere, so bounded since periodic. Furthermore f is purely quadratic, namely

$$f(u+w) + f(u-w) = 2f(u) + 2f(w)\,,$$

and this is then true for all u, w. Then $f(0) = 0$, and

$$f(2u) = 4f(u), \quad \text{so} \quad f(2^n u) = 4^n f(u)\,,$$

whence $f(u) = f(2^n u)/4^n$, and $f(u) = 0$ because f is bounded. This proves the theorem.

Note. The function λ is the "quasi-function" of Néron [Ne 2] associated with the divisor on the elliptic curve consisting of the single point equal to the origin. Néron develops the theory of quasi functions on arbitrary varieties, and especially abelian varieties. The above characterization, specializing Néron's results to elliptic curves, is also valid in the non-archimedean case, cf. Chapter III, and was formulated by Tate.

To get the local Néron function for an arbitrary elliptic curve, let

$$f: A \to B$$

be an isomorphism defined over the complex numbers. The uniqueness shows that

$$\lambda_B \circ f = \lambda_A .$$

In fact, suppose that B is given in Weierstrass form

$$y^2 = x^3 + ax + b ,$$

and we make an isomorphism using multiplication by c, so that

$$x' = c^2 x, \qquad y' = c^3 y, \ldots, \qquad \Delta' = c^{12} \Delta .$$

The expression

$$-\log |x(P) - x(Q)| + \tfrac{1}{6} \log |\Delta|$$

is invariant under this isomorphism, because $\log |c^2|$ cancels. Hence to each given A we find the elliptic curve with the same j-invariant, and corresponding τ in the fundamental domain. The formulas of this section and the next apply to B, and give the Néron function on A by pull back.

§ 8. q-Expansions and Products

We do not derive the q-expansion for the sigma function, but quote it (cf. [L 2] Chapter 18, § 2):

$$\sigma(z; \tau) = (2\pi i)^{-1} e^{\frac{1}{2}\eta_2 z^2} (q_z^{\frac{1}{2}} - q_z^{-\frac{1}{2}}) \prod_{n=1}^{\infty} \frac{(1 - q_\tau^n q_z)(1 - q_\tau^n / q_z)}{(1 - q_\tau^n)^2}$$

where

$$q_\tau = q = e^{2\pi i \tau} \quad \text{and} \quad q_z = e^{2\pi i z}.$$

The complex number η_2 is that associated with the lattice $[\tau, 1]$, with

$$\omega_1 = \tau, \qquad \omega_2 = 1.$$

We consider only this normalized lattice, so that

$$u = u_1\tau + u_2 .$$

Then the Legendre relation immediately yields

$$e^{-\frac{1}{2}\eta(u)u}\, e^{\frac{1}{2}\eta_2 u^2} = q_u^{\frac{1}{2}u_1}.$$

We also recall that

$$\Delta = (2\pi i)^{12} q \prod_{n=1}^{\infty} (1 - q^n)^{24},$$

and we shall meet the Bernoulli polynomial

$$B_2(u_1) = u_1^2 - u_1 + \tfrac{1}{6} .$$

We find:

K 3. $$\mathfrak{k}(u_1, u_2; \tau) = \frac{1}{2\pi i} q_u^{\frac{1}{2}u_1}(q_u^{\frac{1}{2}} - q_u^{-\frac{1}{2}}) \prod \frac{(1 - q^n q_u)(1 - q^n/q_u)}{(1 - q^n)^2} .$$

Multiplying by $\Delta^{1/12}$ yields the q-product for $g(u)$, namely:

$$g(u) = g(u; \tau) = \Delta(\tau)^{1/12}\mathfrak{k}(u; \tau) .$$

K 4. $$g(u; \tau) = q^{\frac{1}{2}B_2(u_1)}\, e^{2\pi i u_2(u_1 - 1)/2}(q_u - 1) \prod_{n=1}^{\infty} (1 - q^n q_u)(1 - q^n/q_u) .$$

It is now convenient to put $t = q_u$ and

$$g_0(t) = g_0(q, t) = (t - 1) \prod_{n=1} (1 - q^n t)(1 - q^n/t) .$$

Then

$$\boxed{g(u) = q^{\frac{1}{2}B_2(u_1)}\, e^{2\pi i u_2(u_1 - 1)/2} g_0(q_u) .}$$

From the q-product we obtain an explicit formula for the Néron function in terms of q and t.

Theorem 8.1. *For an elliptic curve with lattice* $[\tau, 1]$ *we have*

$$\lambda(u) = \tfrac{1}{2}B_2(u_1)v(q) + v(g_0(q_u)) .$$

Note that the invariance properties of the preceding sections show that the function $\lambda = \lambda_L$ is invariant under isomorphisms of the complex torus, i.e. depends only on the isomorphism class. It can be computed by selecting the dehomogenized form of the lattice $[\tau, 1]$ giving rise to the q-product as above.

Define

$$f(u) = f(u;\tau) = (q_u^{\frac{1}{2}} - q_u^{-\frac{1}{2}}) \prod_{n=1}^{\infty} (1 - q^n q_u)(1 - q^n/q_u) .$$

Then

$$f(u;\tau) = e^{-\frac{1}{2}\eta_2 u^2}(\Delta/q)^{1/12}\sigma(u;\tau) .$$

From the expression of the Weierstrass function by the sigma function and the pure quadraticity of the exponent u^2, we find:

Theorem 8.2. *If* u, u' *are complex numbers such that* $q_u, q_{u'}$ *are not integral powers of* q, *then*

$$\wp(u) - \wp(u') = -\frac{f(u+u')f(u-u')}{f(u)^2 f(u')^2}(\Delta/q)^{\frac{1}{6}} .$$

It is convenient to reformulate this formula. We put

$$t = e^{2\pi i u}, \qquad x(t) = \wp(u)$$

$$\psi(t) = f(u) = t^{-\frac{1}{2}}(t-1) \prod_{n=1}^{\infty} (1 - q^n t)(1 - q^n/t) .$$

Then

$$x(t) - x(t') = -\frac{\psi(tt')\psi(t/t')}{\psi(t)^2\psi(t')^2}(\Delta/q)^{\frac{1}{6}}$$

is valid for $t, t' \neq q^n$ for all integers n.

Observe that even though $t^{\frac{1}{2}}$ occurs in the definition of ψ, nevertheless the half power disappears in the formula for the difference

$$x(t) - x(t') .$$

In fact, the formula amounts to an identity in power series in q, with coefficients which are rational functions in t, t'. This is seen from the q-expansion for the Weierstrass function, e.g. as in [L 2], Chapter 4, § 2. The formula then remains valid when the variables take on special values in a field complete under a non-archimedean absolute value, a situation which is discussed in Chapter III, § 5, under conditions of convergence, namely $|q| < 1$.

Next we wish to give an estimate for the difference between the Néron function and the local height, defined by

$$h_v(P) = \log \max \{1, |x(P)|\}$$

where v here denotes the ordinary absolute value on the complex numbers.

We recall the q-expansion for the Weierstrass function, as for instance in [L 2], Chapter 4, § 2. We consider the normalized case corresponding to a lattice $[\tau, 1]$, with τ in the usual fundamental domain, so that in particular,

$$(*) \qquad \operatorname{Im} \tau \geqslant \sqrt{3/2}\,, \quad \text{and} \quad |q| \leqslant e^{-\pi\sqrt{3}}.$$

We have:

$$\frac{1}{(2\pi i)^2} \wp(u; \tau) = \frac{1}{12} + \frac{q_u}{(1-q_u)^2} + \sum_{m=1}^{\infty} \sum_{n=1}^{\infty} n q^{mn} (q_u^n + q_u^{-n}) - 2 \sum_{m=1}^{\infty} \sum_{n=1}^{\infty} n q^{mn}.$$

This is valid for

$$|q| < |q_u| < |q|^{-1}.$$

In view of the periodicity of $\wp$, its value $\wp(u; \tau)$ can always be determined from u satisfying these inequalities. Furthermore, since $\wp$ is an even function, we can select u_1, u_2 such that

$$(**) \qquad u = u_1 \tau + u_2\,, \qquad 0 \leqslant u_1 \leqslant \tfrac{1}{2}\,, \qquad 0 \leqslant u_2 < 1\,.$$

We then have

$$|q^{\frac{1}{2}}| \leqslant |q_u| \leqslant 1\,,$$

whence $|qq_u|$ and $|q/q_u| \leqslant |q|^{\frac{1}{2}}$. Therefore the double series in the expression for the $\wp$-function is uniformly bounded for all values of τ, u subject to conditions (*) and (**). This yields:

Lemma 1. *Assume the normalizations* (*) *and* (**) *on* τ *and* u_1. *Then*

$$\frac{1}{(2\pi i)^2} \wp(u; \tau) = \frac{q_u}{(1-q_u)^2} + O(1)\,,$$

where $O(1)$ *is absolute, independent of* τ, u *subject to* (*), (**).

The only possibility for $\wp(u; \tau)$ to be unbounded lies in the term

$$\frac{q_u}{(1-q_u)^2}\,,$$

which becomes large only when q_u tends to 1. Thus we find:

Lemma 2. *Again under* (*) *and* (**), *we have*

$$\max\{0, \log|\wp(u;\tau)|\} = -2\log|1-q_u| + O(1).$$

From Theorem 8.1 and the bounds for q, q_u we also find

$$\lambda(u) = -\tfrac{1}{2}B_2(u_1)\log|q| - \log|1-q_u| + O(1),$$

$$= -\tfrac{1}{2}(u_1^2 - u_1 + \tfrac{1}{6})\log|q| - \log|1-q_u| + O(1).$$

Theorem 8.3. *Under conditions* (*), (**),

$$\lambda(u) - \tfrac{1}{2}\max\{0, \log|\wp(u;\tau)|\} = -\tfrac{1}{2}B_2(u_1)\log|q| + O(1).$$

$$= \tfrac{1}{2}B_2(u_1)v(q) + O(1).$$

In particular, on the elliptic curve satisfying (*) *we have the estimate*

$$-O(1) - \tfrac{1}{24}v(q) \leqslant \lambda(P) - \tfrac{1}{2}h_v(P) \leqslant \tfrac{1}{12}v(q) + O(1),$$

where the $O(1)$ *is absolute, and similarly with* $v(\Delta)$ *replacing* $v(q)$.

Proof. This is immediate from the above, and from the trivial estimate for the Bernoulli polynomial, for $0 \leqslant u_1 \leqslant \frac{1}{2}$, namely

$$-\tfrac{1}{24} \leqslant \tfrac{1}{2}B_2(u_1) \leqslant \tfrac{1}{12}.$$

Thus the theorem gives an effective bound between the Néron function and the local height function.

Recall the formula

$$\Delta = (2\pi i)^{12} q \prod (1-q^n)^{24}.$$

The product $\prod(1-q^n)^{24}$ is bounded in absolute value away from 0 and ∞. Therefore

$$|\Delta| \gg\ll |q|,$$

and

$$\log|q| = \log|\Delta| + O(1).$$

Thus the estimate in Theorem 8.3 could also be expressed in terms of $v(\Delta)$ instead of $v(q)$.

For an arbitrary elliptic curve we shall analyse the situation by considering an isomorphism with a curve satisfying the previous normalization. Let A be an elliptic curve over the complex numbers, with invariant j. We shall distinguish cases. We let A be isomorphic to A', where A' has period lattice $[\tau, 1]$ with τ in the fundamental domain, and $q = e^{2\pi i \tau}$. Then

$$j = \frac{1}{q} + 744 + 196884q + \cdots .$$

There exists $\delta > 0$ such that if $|q| < \delta$ then

$$j = \frac{1}{q} + O(1) \quad \text{and} \quad |j| \gg\ll \frac{1}{|q|} \gg\ll |\Delta'| .$$

We distinguish:

Case 1. $|q| \geqslant \delta$, corresponding to $|j|$ bounded.

Case 2. $|q| < \delta$, corresponding to $C_0 \leqslant |j|$ for some $C_0 > 1$.

The constants C_0 and $O(1)$ are absolute.

Theorem 8.4. *Let A be as above an elliptic curve over the complex numbers. Then:*

Case 1. $$|\lambda - \tfrac{1}{2}h_v| \leqslant \tfrac{1}{12}|v(\Delta)| + O(1)$$

Case 2. $$|\lambda - \tfrac{1}{2}h_v| \leqslant \tfrac{1}{12}|v(\Delta)| + \tfrac{1}{6}|v(j)| + O(1) .$$

Proof. Let P' be the point on A' corresponding to P on A. There is a number c such that

$$c^{12}\Delta' = \Delta , \qquad c^2 x' = x .$$

We write

$$\lambda(P) - \tfrac{1}{2}h_v(P) = \lambda'(P') - \tfrac{1}{2}h'_v(P') + \tfrac{1}{2}[h'_v(P') - h_v(P)] .$$

We shall use Theorem 8.3 to estimate the first term on the right, and we shall estimate the second term directly. We distinguish cases.

Case 1. $|q| \geqslant \delta$. Then

$$1 > |q| \gg\ll |\Delta'| \gg \delta ,$$

so in this case, $|v(\Delta')| = O(1)$. By Theorem 8.3 we conclude that

$$\lambda'(P') - \tfrac{1}{2}h'_v(P') = O(1) .$$

The second term is equal to

$$\log \max \{1, |x'|\} - \log \max \{1, |c^2 x'|\}$$

so

$$\tfrac{1}{2}|h'_v(P') - h_v(P)| \leqslant |v(c)| = \tfrac{1}{12}|v(\Delta) - v(\Delta')|$$

$$\leqslant \tfrac{1}{12}v(\Delta) + O(1)\,.$$

This concludes Case 1.

Case 2. $|q| < \delta$. Then

$$|j| \gg\ll \frac{1}{|q|} \gg\ll \frac{1}{|\Delta'|} \quad \text{and} \quad |c^{12}| \gg\ll |j\Delta|\,.$$

By Theorem 8.3,

$$|\lambda'(P') - \tfrac{1}{2}h'_v(P')| \leqslant \tfrac{1}{12}|v(\Delta')| + O(1) = \tfrac{1}{12}|v(j)| + O(1).$$

On the other hand,

$$\tfrac{1}{2}|h'_v(P') - h_v(P)| \leqslant |v(c)| = \tfrac{1}{12}|v(j\Delta)| + O(1)\,.$$

This proves the theorem.

Remark. Assume that

$$|\Delta| \gg 1 \quad \text{so} \quad v(\Delta) \leqslant O(1)\,.$$

Such a condition is satisfied in applications, when A is defined over the rationals and a, b, Δ are integers, or similarly when these quantities are algebraic integers. In both cases, we then get $|c| \gg 1$, and hence we find an upper inequality

$$\tfrac{1}{2}[h'_v(P') - h_v(P)] \leqslant O(1)\,.$$

In Case 1, this yields

$$-O(1) + \tfrac{1}{12}v(\Delta) \leqslant \lambda(P) - \tfrac{1}{2}h_v(P) \leqslant O(1)\,.$$

In Case 2, this yields

$$-O(1) + \tfrac{1}{8}v(j) + \tfrac{1}{12}v(\Delta) \leqslant \lambda(P) - \tfrac{1}{2}h_v(P) \leqslant -\tfrac{1}{12}v(j) + O(1)\,.$$

Chapter II. The Division Equation

Let A be an elliptic curve defined over a number field K. Let (x, y) be a generic point on A, and let

$$n(x, y) = (x \circ n, y \circ n) .$$

Then $x \circ n$ and $y \circ n$ can be expressed as rational functions in x, y. In this chapter we develop some of the theory of these functions, giving their degrees, and estimates for their coefficients, by means of recursive relations.

These can be interpreted as giving also the division polynomials. Given a point $(\xi, \eta) = Q$ we consider points $P = (x, y)$ such that

$$nP = Q = (\xi, \eta) .$$

Then the coordinates of Q are obtained from those of P by means of the preceding rational functions, and this equation also yields the algebraic equations satisfied by the coordinates of P over the field generated by the coordinates of Q.

Following our general policy, we derive the multiplication and division equations analytically, using the Weierstrass function, and then point out the essentially algebraic nature of the formula thus obtained.

§ 1. The Division Polynomial

If A is an abelian group, we denote by A_n the subgroup of elements $u \in A$ such that $nu = 0$.

We shall prove that there exists an elliptic function f_n for each integer $n \geqslant 1$ such that

$$f_n(z)^2 = n^2 \prod (\wp(z) - \wp(u)) ,$$

where the product is taken over $u \in (\mathbf{C}/L)_n$ and $u \neq 0$. In fact:

For n odd, all factors in the product occur with multiplicity 2, because the two values $\pm u$ are not congruent mod L, and give rise to the same value of $\wp$.

For n even, all factors occur with multiplicity 2 except those for which $2u \equiv 0 \pmod{L}$, in other words,

$$u \equiv \frac{\omega_1}{2}, \frac{\omega_2}{2}, \frac{\omega_3}{2} \qquad \text{(where } \omega_3 = \omega_1 + \omega_2\text{)},$$

and these have multiplicity 1. At these points $\wp - \wp(u)$ has a double zero, and

$$\wp'^2 = 4 \prod_{\substack{2u=0 \\ u \neq 0}} (\wp - \wp(u)) .$$

Therefore the product defining f_n^2 is a perfect square, and we can pick f_n with the appropriate sign such that:

(1) **n odd:** $f_n = P_n(\wp)$, where P_n is a polynomial of degree $(n^2 - 1)/2$, leading coefficient n,

$$f_n = n\wp^{(n^2-1)/2} + \cdots$$

(2) **n even:** $f_n = \frac{1}{2}\wp' P_n(\wp)$, where P_n is a polynomial of degree $(n^2 - 4)/2$, leading coefficient n,

$$f_n = \frac{n}{2}\wp'\wp^{(n^2-4)/2} + \cdots .$$

In all cases, we may therefore write the expansion of f_n at $z = 0$ in the form

$$f_n(z) = \frac{(-1)^{n+1} n}{z^{n^2-1}} + \cdots . \tag{3}$$

We let

$$\wp_n(z) = \wp(nz) .$$

Theorem 1.1. *We have*

$$\wp_n = \wp - \frac{f_{n+1} f_{n-1}}{f_n^2} . \tag{4}$$

Proof. Except for the lattice points, we note that

$$\wp(nz) - \wp(z)$$

(i) has poles exactly at the zeros of f_n^2, with the same multiplicity 2, because $z \mapsto nz$ is a local analytic isomorphism;

(ii) has zeros at those points z such that $nz \equiv \pm z \pmod{L}$, i.e. at those z such that

$$(n+1)z \quad \text{and} \quad (n-1)z \equiv 0 \pmod{L}.$$

These have multiplicity 1, because taking the derivative shows that we cannot have

$$n\wp'(nz) = \wp'(z)$$

if $nz \equiv \pm z \pmod{L}$. Hence the function

$$\frac{f_n^2(\wp_n - \wp)}{f_{n+1}f_{n-1}}$$

is elliptic, and has no zeros or poles other than at 0 on the torus. It is therefore constant. At 0, its expansion is

$$\frac{n^2(-n^2+1)/n^2}{(n+1)(n-1)} = -1.$$

This proves the theorem.

We shall now compute $\wp_2$ and $\wp_3$ explicitly. The addition formula shows that

$$\wp_2 = -2\wp + \tfrac{1}{4}(\wp''/\wp')^2.$$

But $\wp'' = 6\wp^2 - \frac{1}{2}g_2$ and hence

(5) $$\wp_2 = \wp - \frac{3\wp^4 - \frac{3}{2}g_2\wp^2 - 3g_3\wp - \frac{1}{16}g_2^2}{\wp'^2}.$$

Let $f_1 = 1$, $f_2 = \wp' = 2\wp'/2$. Then from

$$\wp_2 - \wp_1 = -f_3/\wp'^2$$

we get

(6) $$\boxed{f_3 = 3\wp^4 - \tfrac{3}{2}g_2\wp^2 - 3_{g_3}\wp - \tfrac{1}{16}g_2^2.}$$

To get f_4, we use the formulas

$$\wp(u+v) = -\wp u - \wp v + \frac{1}{4}\left(\frac{\wp'u - \wp'v}{\wp u - \wp v}\right)^2$$

$$\wp(u-v) = -\wp u - \wp v + \frac{1}{4}\left(\frac{\wp'u + \wp'v}{\wp u - \wp v}\right)^2$$

whence subtracting yields

$$\wp(u+v) - \wp(u-v) = -\frac{\wp' u \wp' v}{(\wp u - \wp v)^2}.$$

We put $u = 2v$, and $v = z$ to get

$$\begin{aligned}\wp_3 - \wp &= -\frac{\wp'(2z)\wp'(z)}{(\wp_2 - \wp_1)^2} \\ &= -\frac{\wp'(2z)\wp'^5}{f_3^2} \\ &= -\frac{f_4 f_2}{f_3^2}.\end{aligned}$$

Differentiating $\wp(2z)$ in (5), we replace $\wp'(2z)$ by a polynomial in $\wp$, $\wp'$, and solve for f_4 in

$$f_4 f_2 = \wp'(2z)\wp'^5.$$

This yields:

$$\boxed{f_4 = \tfrac{1}{2}\wp'(4\wp^6 - 5g_2\wp^4 - 20g_3\wp^3 - \tfrac{5}{4}g_2^2\wp^2 - g_2g_3\wp - 2g_3^2 + \tfrac{1}{16}g_2^3).}$$

To compute further f_n we give a recursion formula.

Theorem 1.2. *Let $m > n$. Then*

$$f_{m+1}f_{m-1}f_n^2 - f_{n+1}f_{n-1}f_m^2 = f_{m+n}f_{m-n}.$$

Proof. From the formula

$$\wp - \wp_n = \frac{f_{n+1}f_{n-1}}{f_n^2} \quad \text{and} \quad \wp - \wp_m = \frac{f_{m+1}f_{m-1}}{f_m^2}$$

we see that $\wp(mz) - \wp(nz)$ has a zero at those u such that

$$mu \equiv \pm nu \not\equiv 0 \pmod{L},$$

i.e. $(m \pm n)u \equiv 0 \pmod{L}$, of multiplicity 1 (differentiate and note that

$$m\wp'(mu) - n\wp'(nu) \neq 0).$$

But f_n, f_m cannot have a zero at these points, because $mu, nu \not\equiv 0$. Hence these points are the zeros of

$$f_{n+1}f_{n-1}f_m^2 - f_{m+1}f_{m-1}f_n^2 .$$

But $f_{n+m}f_{m-n}$ has the same zeros, and both above functions are polynomials in $\wp$, so have a pole only at 0 (mod L). Hence they are constant multiples of each other. The expansion at 0 gives the constant 1 for their quotient, thereby proving the theorem.

We put $(m, n) = (n + 1, n)$ or $(n + 1, n - 1)$. We then obtain the recursion formulas:

Theorem 1.3. $f_{2n+1} = f_{n+2}f_n^3 - f_{n+1}^3 f_{n-1}$

$$\wp' f_{2n} = f_{2n}f_2 = f_n(f_{n+2}f_{n-1}^2 - f_{n-2}f_{n+1}^2) .$$

§ 2. The Algebraic Formulas Over Z

It is convenient to renormalize the equation of the elliptic curve so as to end up with coefficients for the division equation which are polynomials over the integers. Thus we let

$$x = \wp , \qquad y = \tfrac{1}{2}\wp' , \qquad a = -\tfrac{1}{4}g_2 , \qquad b = -\tfrac{1}{4}g_3 .$$

Then the elliptic curve can be written in the form

$$y^2 = x^3 + ax + b .$$

Furthermore,

$$f_1 = 1 , \qquad f_2 = 2y$$

$$f_3 = 3x^4 + 6ax^2 + 12bx - a^2$$

$$f_4 = 4y(x^6 + 5ax^4 + 20bx^3 - 5a^2x^2 - 4abx - 8b^2 - a^3) .$$

We let

$$f_n = \psi_n(x, y)$$

and write

$$x \circ n = \wp_n .$$

The above formulas show that for $n = 1, 2, 3, 4$ we can write

$$f_n = P_n(x) \qquad \text{for } n \text{ odd}$$

$$f_n = 2yP_n(x) \quad \text{for } n \text{ even} ,$$

where $P_n(x)$ is in $\mathbf{Z}[a, b, x]$. From Theorem 1.3 we find inductively that this is true for all n, because we get the inductive relations:

(1) $$P_{2n+1} = f_{2n+1} = \begin{cases} P_{n+2}P_n^3 - P_{n+1}^3 P_{n-1} \cdot 16y^4 & n \text{ odd} \\ 16y^4 P_{n+2}P_n^3 - P_{n+1}^3 P_{n-1} & n \text{ even}. \end{cases}$$

$$P_{2n} = P_n(P_{n+2}P_{n-1}^2 - P_{n-2}P_{n+1}^2)$$

because $(2y)^2$ cancels on both sides for this last formula, whether n is odd or even.

In particular, we get from (1) the value of the constant term:

(2) $$\psi_{2n+1}(0) = (-1)^n a^{n^2+n} \quad \text{if } b = 0.$$

Theorem 2.1. *Let*

$$\varphi_n(x) = x\psi_n^2 - \psi_{n+1}\psi_{n-1}$$

$$4y\omega_n = \psi_{n+2}\psi_{n-1}^2 - \psi_{n-2}\psi_{n+1}^2.$$

Then

(i) $$n(x, y) = \left(\frac{\varphi_n}{\psi_n^2}, \frac{\omega_n}{\psi_n^3}\right).$$

(ii) *The expressions* φ_n, ψ_n *(for n odd) and* $\psi_n/2y$ *(for n even) are polynomials in* $\mathbf{Z}[a, b, x]$. *We have*

$$\varphi_n(x) = x^{n^2} + \cdots \quad \textit{with leading coefficient } 1$$

$$\psi_n^2(x) = n^2 x^{n^2-1} + \cdots \quad \textit{with leading coefficient } n^2.$$

(iii) $y^{-1}\omega_n$ (for n odd) *and* ω_n *(for n even) is a polynomial in* $\mathbf{Z}[a, b, x]$.

This merely summarizes the preceding discussion.

We also get the following divisibility properties of Cassels [Ca 2].

Theorem 2.2.

(i) *Let* 2^s *divide n exactly. Then* 2^{2s} *is the g.c.d. of the coefficients of* $\psi_m^2(x, a, b) \in \mathbf{Z}[x, a, b]$.

(ii) *If* $n = 2^s$ *then*

$$2^{-2s}\psi_n^2 = x^{n^2-1} + \textit{a polynomial in } \mathbf{Z}[x, a, b] \textit{ of degree} < n^2 - 1 \textit{ in } x.$$

(iii) *If $n = p^s$ with an odd prime p, then*

$$g_{p^s}(x) = \psi_{p^s}^2/\psi_{p^{s-1}}^2$$

lies in $\mathbf{Z}[x, a, b]$, has leading coefficient p^2, and relatively prime coefficients in $\mathbf{Z}$.

Proof. Suppose n is odd. Then the value of the constant term in (2) shows that the coefficients of $\psi_n(x, a, b)$ are relatively prime. The assertion (i) then follows for n even by induction. Assertion (ii) then is a consequence of (ii) in the preceding theorem. Finally (iii) follows by induction on s from (i) and Gauss' lemma.

In all the above formulas, the expressions are polynomials in x, y, a, b and the analysis has disappeared. Furthermore, these expressions are homogeneous in the following sense.

Suppose that a, b are algebraically independent, but x, y are related by the given equation. We can give a grading to the ring

$$\mathbf{Z}[a, b, x, y]$$

by ascribing the following weights:

a has weight 4

b has weight 6

x has weight 2

y has weight 3.

A monomial $a^{n_1}b^{n_2}x^{n_3}y^{n_4}$ has weight $4n_1 + 6n_2 + 2n_3 + 3n_4$. Then the polynomials φ_n, ψ_n, P_n are all homogeneous with respect to this grading. For instance, ψ_n has weight n.

The reader should also get used to the idea that all the formulas are valid for any elliptic curve in arbitrary characteristic $\neq 2, 3$. Indeed, we may start with the equation

$$y^2 = x^3 + ax + b,$$

and define an addition on the set of solutions of this equation, by the same formulas as in the analytic case. If

$$(x_1, y_1) + (x_2, y_2) = (x_3, y_3)$$

then we let

$$(3) \qquad x_3 = -x_2 - x_1 + \left(\frac{y_2 - y_1}{x_2 - x_1}\right)^2$$

$$= \frac{a(x_1 + x_2) + 2b + x_1x_2^2 + x_2x_1^2 - 2y_1y_2}{(x_1 - x_2)^2}.$$

We put

$$y_3 = -(\lambda x_3 + \mu) \tag{4}$$

where

$$\lambda = \frac{y_2 - y_1}{x_2 - x_1} \quad \text{and} \quad \mu = \frac{x_2 y_1 - x_1 y_2}{x_2 - x_1}. \tag{5}$$

These formulas are of course for the case when $x_1 \neq x_2$. We may view x, y as functions of the points on the curve. For the duplication formula, if P is an arbitrary point, then we have to make the special definition

$$\begin{aligned} x(2P) &= -2x + \left(\frac{3x^2 + a}{2y}\right)^2 \\ &= \frac{x^4 - 2ax^2 - 8xb + a^2}{4(x^3 + ax + b)}. \end{aligned} \tag{6}$$

It takes little imagination to believe that the set of points on the curve, together with another point called infinity as the origin, form an abelian group under this addition law. It takes even less imagination to believe that a brute force algebraic verification is a pain.

However, let us assume that the coefficients a, b are in a complete discrete valuation ring R of characteristic O, and let $\bar{R}$ be the residue class field. (The bar denotes reduction mod the maximal ideal). Let Δ be the discriminant, and assume $\bar{\Delta} \neq 0$. Assume also that the characteristic of $\bar{R}$ is $\neq 2, 3$. If $P = (x, y)$ is a point of A in the quotient field K of R, we define:

$$\begin{aligned} \bar{P} &= \bar{O} && \text{if } x \notin R \\ \bar{P} &= (\bar{x}, \bar{y}) && \text{if } x \in R\,. \end{aligned}$$

The points $A(K)$ form a group whose group law is given by (3) for the x-coordinates. It is then rather simple to verify that the map

$$P \mapsto \bar{P}$$

is a homomorphism, if we define addition on $\bar{A}_K$ by the same formula (putting bars on the coordinates). Thus we define addition on the curve $\bar{A}$, with equation

$$\bar{y}^2 = \bar{x}^3 + \bar{a}\bar{x} + \bar{b}$$

by reduction of addition on A. In this way one can handle elliptic curves in characteristic >0 (up to a point) without any major foundational difficulties.

The reason why the reduction process works is that the formulas arising from analysis turn out to involve only the coordinates x, y and a, b with integer coefficients.

When studying the quadraticity of the height, we shall also use the following formulas, which we list here for completeness of reference. We let

$$(x_1, y_1) - (x_2, y_2) = (x_4, y_4).$$

Then:

$$x_3 + x_4 = \frac{2(x_1 + x_2)(a + x_1 x_2) + 4b}{(x_1 - x_2)^2} \tag{7}$$

$$x_3 x_4 = \frac{(x_1 x_2 - a)^2 - 4b(x_1 + x_2)}{(x_1 - x_2)^2}. \tag{8}$$

These are easily proved using the Weierstrass function, and the proofs are left to the reader.

Observe that when we put $x_1 = x_2$ in the numerator of formula (8) for $x_3 x_4$, then this numerator becomes exactly the numerator of $x(2P)$, namely

$$\boxed{\varphi_2(x) = x^4 - 2ax^2 - 8xb + a^2 = -8xy^2 + (3x^2 + a)^2.}$$

Theorem 2.3. *The polynomials $\varphi_n(x)$ and $\psi_n^2(x)$ are relatively prime to each other.*

Proof. In view of the expression of $x \circ n$ as a quotient of polynomials,

$$x \circ n = \varphi_n(x)/\psi_n^2(x),$$

we see that x satisfies an algebraic equation

$$\psi_n^2(x) x \circ n - \varphi_n(x) = 0$$

with respect to $x \circ n$. If K is a field containing the coefficients a, b of the curve, we see that φ_n, ψ_n^2 being relatively prime is equivalent to x being of degree n^2 over $K(x \circ n)$. One can give an algebraic proof for this fact which is not too difficult, but we follow our policy of looking at things over the complex numbers, where we put $x = \wp$, $y = \wp'/2$. In this case,

$$\mathbf{C}(\wp, \wp') \textit{ is an extension of degree } n^2 \text{ over } \mathbf{C}(\wp_n, \wp'_n)$$

because the latter field is exactly the fixed field of $\mathbf{C}(\wp, \wp')$ under the group of automorphisms obtained by translations,

$$z \mapsto z + t,$$

where $t \in (\mathbf{C}/L)_n$, which has order n^2. We may then consider the following diagram of fields.

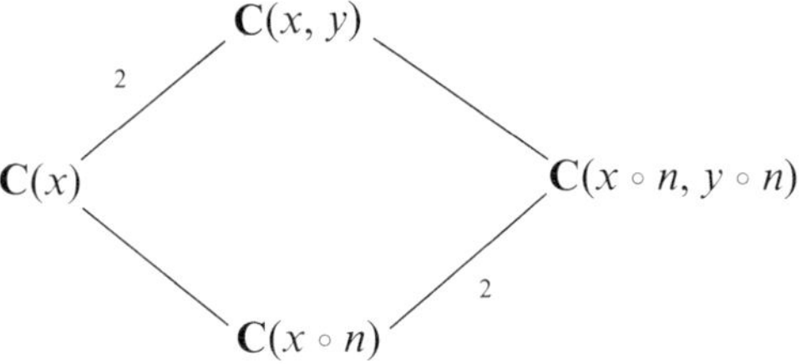

We must have

$$[\mathbf{C}(x, y) : \mathbf{C}(x)] = [\mathbf{C}(x \circ n, y \circ n) : \mathbf{C}(x \circ n)]$$

because y and $y \circ n$ are odd functions, and thus $\mathbf{C}(x, y)$ is the composite of $\mathbf{C}(x)$ and $\mathbf{C}(x \circ n, y \circ n)$. Hence

$$[\mathbf{C}(x) : \mathbf{C}(x \circ n)] = [\mathbf{C}(x, y) : \mathbf{C}(x \circ n, y \circ n)] = n^2 .$$

This proves the theorem.

The argument also works purely algebraically, replacing $\mathbf{C}$ by a field K of characteristic $\neq 2, 3$, containing a, b, provided we know that $K(x, y)$ has degree n^2 over $K(x \circ n, y \circ n)$. Again, the Galois theory argument works equally well in this case, provided we know that the elliptic curve has precisely n^2 distinct points of order n, which is true if the characteristic does not divide n. Let us denote the group of points of order n on the elliptic curve A by A_n. We take these points in some fixed algebraically closed field containing K, and also containing a point $(x, y) = X$ where x, y are transcendental over K, satisfying the equation of the curve,

$$y^2 = x^3 + ax + b .$$

We assume that K contains all the coordinates of all points in A_n. For each $Q \in A_n$ there is an automorphism τ_Q of $K(x, y) = K(X)$ such that

$$\tau_Q(X) = X + Q .$$

These automorphisms form a group isomorphic to A_n, and called the group of translations of $K(X)$ by A_n. Its fixed field is precisely

$$K(x \circ n, y \circ n) .$$

These assertions are obvious when K is, say, a finite extension of the rational numbers containing all the coordinates of points Q in A_n, and when $x = \wp$, $y = \wp'/2$. The reader should get used to both notations, involving on the one hand the parametrization by the Weierstrass functions, giving the analytic interpretation, and on the other hand, the purely algebraic formulation in terms of the algebraic coordinate functions x, y.

Theorem 2.3 is of course relative to the constant *field*. It is of some interest to see integrally the extent to which the polynomials φ_n and ψ_n^2 have a greatest common divisor. The most interesting special case is for $n = 2$.

Theorem 2.4. *Let* $\Delta_0 = 4a^3 + 27b^2$. *Then*

$$(3x^2 + 4a)\varphi_2(x) - (3x^3 - 5ax - 27b)y^2 = \Delta_0 .$$

Proof. We eliminate x between the equations

$$\varphi_2(x) = x^4 - 2ax^2 - 8bx + a^2$$
$$y^2 = x^3 + ax + b ,$$

and the assertion drops out.

§ 3. Estimates for the Coefficients

Theorem 3.1. *The polynomials* $\varphi_n(x)$ *and* $\psi_n^2(x)$ *have coefficients bounded in absolute value by*

$$C^{n^2},$$

where C *is a constant depending only on* a, b.

Proof. It is possible to give an easy inductive argument for this estimate, using the recursion relations giving the polynomials under consideration. We shall leave this method to the reader, and give a proof by another method based on analysis.

Let T be a variable. Let

$$F(T) = \sum a_i T^i \quad \text{and} \quad G(T) = \sum b_i T^i$$

be polynomials, where the coefficients a_i are complex, and b_i are real $\geqslant 0$. We write

$$F \prec G$$

to mean that $|a_i| \leqslant b_i$ for all i, and say that G **dominates** F. It is clear that the relation of domination is preserved under sums and products.

We note that φ_n and ψ_n^2 can be expressed in terms of f_n^2, f_{n+1}, f_{n-1} as in Theorem 2.1. What we want will then result from the next theorem, giving estimates for the coefficients of f_n as a polynomial in $\wp$, and an easy lemma afterwards.

Theorem 3.2. *There is a constant C depending only on the lattice L such that the coefficients of the polynomial*

$$\prod_{\substack{u \in (\mathbf{C}/L)_n \\ u \neq 0}} (T - \wp u)$$

are bounded by C^{n^2}.

Proof. The polynomial is dominated by

$$\prod_{|\wp u| \geqslant 1} |\wp u| C_1^{n^2} (T+1)^{n^2}.$$

Thus we are reduced to estimating the product

$$\prod_{|\wp u| \geqslant 1} |\wp u|.$$

Let S_1, S_2, S_3, S_4 be the four parallelograms as on the figure, and let S be one of them, say consisting of those division points

$$\frac{r\omega_1 + s\omega_2}{n},$$

with $n/2 \leqslant r \leqslant n$ and $0 \leqslant s \leqslant n/2$.

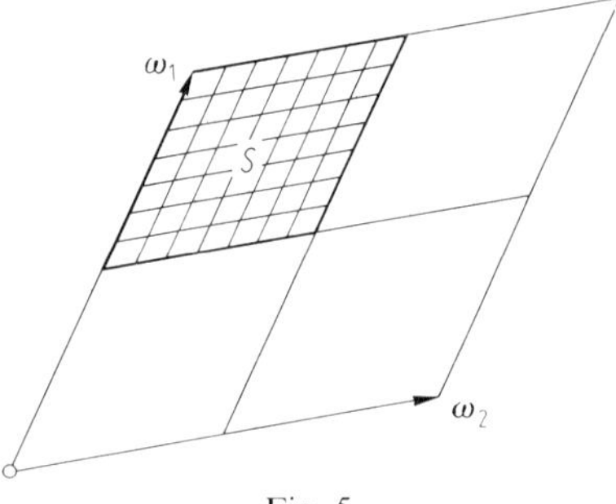

Fig. 5

For

$$u = \frac{r\omega_1 + s\omega_2}{n}$$

we have

$$|\wp u| \ll \frac{1}{\operatorname{dist}(u, L)^2}$$

where dist (u, L) is the distance from u to the lattice, taken with respect to the sup norm, so for the chosen parallelogram,

$$\operatorname{dist}(u, L) = \max\left(\frac{n-r}{n}, \frac{s}{n}\right).$$

Then

$$\begin{aligned}
\prod_{\substack{u \in S \\ |\wp u| \geq 1}} |\wp u| &\leqslant C_2^{n^2} \prod_{\substack{u \in S \\ |\wp u| \geq 1}} \frac{1}{\operatorname{dist}(u, L)^2} \\
&\leqslant C_2^{n^2} \prod_{u \in S} \frac{1}{\operatorname{dist}(u, L)^2} \\
&\leqslant C_2^{n^2} \prod_{m=1}^{[n/2]} \prod_{m/n \leqslant \operatorname{dist}(u, L) \leqslant (m+1)/n} \left(\frac{n}{m}\right)^2 \\
&\leqslant C_3^{n^2} \prod_{m=1}^{n} (n/m)^{2m} \\
&\leqslant C_3^{n^2} \prod_{m=1}^{n} (n/m)^{2n} \\
&\leqslant C_3^{n^2} (n^n/n!)^{2n} \leqslant C_4^{n^2},
\end{aligned}$$

as was to be proved.

The polynomials which we want are essentially factors of the polynomial which we have just estimated. Estimates for the coefficients then follow from a quite general lemma.

Lemma 1. *Let $f(T) \in \mathbf{C}[T]$ be a polynomial of degree d, and let*

$$f(T) = \prod_{i=1}^{d} (T - \alpha_i) = \sum a_i T^i.$$

be a factorization. Let $\|f\| = \max |a_i|$. Then

$$\frac{1}{2^d} \prod_{i=1}^{d} \sup(1, |\alpha_i|) \leqslant \|f\| \leqslant 2^d \prod_{i=1}^{d} \sup(1, |\alpha_i|).$$

Proof. The right inequality is trivially proved by induction, estimating the coefficients in a product of a polynomial $g(T)$ by $(T - \alpha)$. We prove the other by induction on the number of indices i such that $|\alpha_i| > 2$. If $|\alpha_i| \leqslant 2$ for all i, our assertion is obvious. Suppose now that

$$f(T) = g(T)(T - \alpha)$$

with $|\alpha| > 2$, and suppose that our assertion is true for

$$g(T) = T^d + b_{d-1}T^{d-1} + \cdots + b_0 .$$

We have

$$f(T) = T^{d+1} + (b_{d-1} - \alpha)T^d + (b_{d-2} - \alpha b_{d-1})T^{d-1} + \cdots + (-\alpha)b_0 .$$

We can assume $\|g\| = |b_i| \geqslant |b_{i-1}|$ for some i with $0 \leqslant i \leqslant d$ (with the convention $b_d = 1$, $b_{-1} = 0$). Then

$$\begin{aligned} \|f\| \geqslant |\alpha b_i - b_{i-1}| &\geqslant |\alpha||b_i| - |b_{i-1}| \\ &\geqslant |\alpha||b_i| - |b_i| = (|\alpha| - 1)|b_i| \\ &\geqslant \tfrac{1}{2}|\alpha||b_i| = \tfrac{1}{2}|\alpha| \, \|g\| , \end{aligned}$$

and our lemma is obvious since $|\alpha| > 2$.

As a consequence of the lemma, we immediately obtain for two polynomials f, g the inequalities

Lemma 2.
$$\frac{1}{4^d}\|fg\| \leqslant \|f\|\,\|g\| \leqslant 4^d\|fg\| ,$$

by factorizing them into linear factors. Thus a factor of a polynomial has coefficients bounded in the obvious manner by the coefficients of the polynomial. This gives us the desired estimate for factors of the polynomial in Theorem 3.2.

Chapter III. *p*-Adic Addition

In this chapter we investigate addition on the elliptic curve in relation to divisibility properties of the denominators of its coordinates, and the quadraticity of the additional formula. This takes place in the general framework that a Lie group has an exponential map convergent near the origin, and giving one-parameter analytic subgroups. We want to see what happens when the base field is p-adic. As before, we carry out the theory ad hoc in a simple manner, making use of the addition formulas given explicitly on the elliptic curve, without fancy language.

The p-adic study of elliptic curves was originated by Lutz [Lu], see also Weil [We 3]. For the exponential map on abelian varieties or algebraic groups, cf. Mattuck [Mat], Igusa [Ig], and Serre's notes [Se 3]. Cf. also the discussion by Tate in his general report [Ta 1]. The exposition of this chapter also owes much to other notes of Tate, from his Phillips Lectures at Haverford College, and to his article [Ta 2]. For simplicity we usually limit ourselves to curves defined by an equation

$$y^2 = x^3 + ax + b\,,$$

instead of the general equation also valid in characteristic 2 and 3, for which normal forms were originally given by Deuring. By referring to [Ta 1] and [Ta 2], the reader can work such cases out for himself.

Throughout this chapter, we let R be an integral domain with quotient field K, and assume that R is a principal ideal ring. The reader may assume that R has characteristic 0, *but what we say will be true in characteristic* $\neq 2$, 3, *and, suitably formulated is even true in all characteristics.*

The ordinary integers constitute an example of such a ring. If p is a prime number, the local ring $\mathbf{Z}_{(p)}$ consisting of all quotients m/n, where $(n, p) = 1$ has a unique prime element p, and unique factorization in this ring is of the form

$$a = p^r u$$

where u is a unit in $\mathbf{Z}_{(p)}$, i.e. the numerator and denominator of u are not divisible by p. More generally, if π is a prime element of R, then one may similarly form the local ring $R_{(\pi)}$, consisting of all elements a/b, with $a, b \in R$ and b not divisible by π. Unique factorization in $R_{(\pi)}$ is of the form

$$a = \pi^r u ,$$

where u is a unit in $R_{(\pi)}$. We call $r = \operatorname{ord}_\pi a$.

Sections § 4, § 5 and § 6 may be omitted without impairing the logical development of the theory of the height, and of the rest of the book, save for the more refined results.

§ 1. Addition Near the Origin

For this entire section, we assume in addition that R has a single prime element (up to units), which we denote by π. Then R has a unique maximal ideal $(\pi) = R\pi$. We suppose as before that A is defined by the equation

$$y^2 = x^3 + ax + b$$

and assume that $a, b \in R$.

If $x = \pi^{-r}u$ where u is a unit in R, and $r \geqslant 0$, then we say that x has a **pole** of order r (at π). Suppose (x, y) is a point on A with coordinates $x, y \in K$, and x has a pole of order $\geqslant 1$. Then $x^3 + ax + b$ has a pole of order 3 times the order of the pole of x, and consequently y must also have a pole. It then follows that there exist units u, u' in R such that

$$x = \frac{u}{\pi^{2r}} \quad \text{and} \quad y = \frac{u'}{\pi^{3r}}$$

for some integer $r \geqslant 1$. Thus x has a pole of even order, and y has a pole of order divisible by 3.

For $r > 0$ we let $A(\pi^r)$ be the set of points P in $A(K)$ such that P is at infinity, or the denominator of $x(P)$ is divisible by π^{2r}. In this case, this denominator is of the form $\pi^{2r'}$, and the denominator of $y(P)$ is of the form $\pi^{3r'}$.

We let

$$t = x/y \quad \text{and} \quad s = 1/y .$$

In terms of the coordinates t, s the equation for the curve becomes

$$s = t^3 + as^2t + bs^3 .$$

Viewing the curve as embedded in projective plane, the new coordinates are such that the point at infinity in terms of the (x, y) coordinates is transformed to the point $(0, 0)$ in terms of the (t, s) coordinates. One usually calls t a **local parameter** at the origin. We may then characterize $A(\pi^r)$ in terms of these new coordinates as the set of points P such that $t(P)$ is divisible by π^r and $s(P)$ is divisible by π^{3r}.

Let $P_1 = (t_1, s_1)$ and $P_2 = (t_2, s_2)$ be two points in $A(K)$, so that

$$s_2 = t_2^3 + as_2^2 t_2 + bs_2^3$$
$$s_1 = t_1^3 + as_1^2 t_1 + bs_1^3 .$$

Then

$$\begin{aligned} s_2 - s_1 &= t_2^3 - t_1^3 + a(s_2^2 t_2 - s_1^2 t_1) + b(s_2^3 - s_1^3) \\ &= t_2^3 - t_1^3 + a(s_2^2(t_2 - t_1) + (s_2^2 - s_1^2)t_1) + b(s_2^3 - s_1^3) . \end{aligned}$$

Therefore, if $t_2 \neq t_1$ then

$$\frac{s_2 - s_1}{t_2 - t_1}(1 - a(s_1 + s_2)t_1 - b(s_2^2 + s_1 s_2 + s_1^2)] = (t_2^2 + t_1 t_2 + t_1^2) + as_2^2 .$$

Dividing yields an expression for the slope of the line passing through P_1 and P_2. Observe that if $P_3 = (s_3, t_3)$ is the point such that

$$P_1 + P_2 + P_3 = O ,$$

then P_3 is characterized as the third point on the intersection of A with the line

$$s = \alpha t + \beta , \qquad \beta = s_1 - \alpha t_1 ,$$

where α is the **slope**,

$$\alpha = \frac{s_2 - s_1}{t_2 - t_1} .$$

Indeed, this is how addition on the elliptic curve in terms of the (x, y) coordinates came out, and the (s, t) coordinates are obtained merely by "dehomogenizing" the projective equation in terms of another component. Hence this description of the addition law applies equally to the new coordinates. In particular, we obtain

(1) $$\alpha = \frac{t_2^2 + t_1 t_2 + t_1^2 + as_2^2}{1 - a(s_1 + s_2)t_1 - b(s_2^2 + s_1 s_2 + s_1^2)} .$$

This formula is also valid if $t_1 = t_2$.

Theorem 1.1. *The set of points $A(\pi^r)$ is a group. The map*

$$P \mapsto t(P)$$

induces an isomorphism

$$A(\pi^r)/A(\pi^{5r}) \to \pi^r R/\pi^{5r} R .$$

Proof. If t_1, t_2 are divisible by π^r and s_1, s_2 are divisible by π^{3r} then we see that the denominator of α is of the form $1 + z$, where $z \equiv 0 \pmod{\pi}$, and hence is a unit. Furthermore,

$$\alpha \equiv 0 \pmod{\pi^{2r}} .$$

But t_1, t_2, t_3 are the roots of the equation

$$\alpha t + \beta = t^3 + a(\alpha t + \beta)^2 t + b(\alpha t + \beta)^3 ,$$

which, if we rearrange terms, yields

$$0 = t^3(1 + b\alpha^3 + a\alpha^2) + t^2(2a\alpha\beta + 3b\alpha^2\beta) + \cdots.$$

This implies that

$$t_1 + t_2 + t_3 = -\frac{2a\alpha\beta + 3b\alpha^2\beta}{1 + b\alpha^3 + a\alpha^2} . \tag{2}$$

Since $s_1 = \alpha t_1 + \beta$, we conclude that π^{3r} divides β. Hence

$$-t_3 \equiv t_1 + t_2 \pmod{\alpha\beta}$$

and therefore

$$-t_3 \equiv t_1 + t_2 \pmod{\pi^{5r}} .$$

This proves that $A(\pi^r)$ is a group, and the map is an injective homomorphism. We postpone surjectivity to § 3.

Corollary 1. *Let π divide the prime number p. Then there is no torsion prime to p in $A(\pi^r)$ for $r \geqslant 1$.*

Corollary 2. *If n is a positive integer and not a p-power, and $P \in A(K)$ is a point of exact period n, then*

$$x(P) \in R .$$

Proof. Suppose π divides the denominator of x. We write $n = l^m n_0$ where $l \nmid n_0$ and $\pi \nmid l$. Then $n_0 P \neq O$, and has period equal to a power of l, which is prime to π. This contradicts Corollary 1.

Theorem 1.2. *Let $P, Q \in A(K)$ and let n be a positive integer such that $nP = Q$.*

Then

$$\text{den}\, x(P) \mid \text{den}\, x(Q)\,,$$

where den *denotes denominator.*

Proof. This is obvious from the fact that $A(\pi^r)$ is a group.

The preceding theorem shows that the denominators of the division points of a fixed point $Q \neq O$ are bounded.

Theorem 1.3. *Let $P \in A(K)$ be a point of finite order. Assume that π divides the prime number p, and that e is the ramification index, that is $p = \pi^e u$ where u is a unit. If $P \in A(\pi^r)$ with $r \geq 1$ then $r \leq e/4$.*

Proof. By Corollary 2 we may assume that P has period p^m for some positive integer m. Then $t(P) \mid t(p^{m-1}P)$, so it suffices to prove the desired assertion for the point $Q = p^{m-1}P$, which has period p. Let $\pi^r \mid t(Q)$ exactly. Then

$$0 = t(pQ) \equiv pt(Q) \bmod \pi^{5r}\,,$$

so that π^{5r} divides π^{e+r}. This yields $4r \leq e$, and proves the theorem.

Remark. Occasionally, it is convenient to normalize the Weierstrass form somewhat less stringently than we have done, and to write the equation for the elliptic curve in the form

$$y^2 = x^3 + ax^2 + bx + c\,,$$

where $a, b, c \in R$. We let $t = x/y$ and $s = 1/y$ as before, and define $A(\pi^r)$ as before. It is still true that

$$P_1 + P_2 + P_3 = O$$

if and only if the three points lie on the same line. The slope α of the line passing through the three points is given in the present case by the formula

$$\alpha = \frac{t_2^2 + t_1 t_2 + t_1^2 + as_1(t_1 + t_2) + bs_1^2}{1 - at_2^2 - bt_2(s_1 + s_2) - c(s_1^2 + s_1 s_2 + s_2^2)}$$

and

$$t_1 + t_2 + t_3 = -\frac{a\beta + 2b\alpha\beta + 3c\alpha^2\beta}{1 + a\alpha + b\alpha^2 + c\alpha^3}\,.$$

In this case, note the presence of $a\beta$ in the numerator. This yields the weaker congruence than before,

$$-t_3 \equiv t_1 + t_2 \pmod{\pi^{3r}}$$

if $P_1, P_2 \in A(\pi^r)$. It suffices to give results analogous to the previous ones, with $3r$ replacing $5r$, so that

$$A(\pi^r)/A(\pi^{3r}) \approx \pi^r R/\pi^{3r} R\,.$$

The estimate for r in Theorem 1.3 is then replaced by $r \leqslant e/2$.

A similar analysis works for the most general normal forms, cf. [Ta 2], Theorem 4.2 et seq.

We may also obtain bounds for torsion points, following Lutz–Nagell [Lu] and [Na], and especially Cassels [Ca 1], Theorem 17.2, or [Ca 2]. Recall that the discriminant of $x^3 + ax + b$ is $-16(4a^3 + 27b^2)$.

Theorem 1.4. *Let $P \in A(K)$, $P = (x, y)$ and let*

$$2P = (x_2, y_2)\,.$$

Let $\Delta_0 = 4a^3 + 27b^2$. Then

$$y^2(4x_2(3x^2 + 4a) - 3x^3 + 5ax + 27b) = \Delta_0\,.$$

In particular, if both P and $2P$ are integral points (that is, have coordinates in R), then $y^2 \mid \Delta_0$.

Proof. We have

$$x_2 = \frac{\varphi_2(x)}{4y^2}$$

where

$$\varphi_2(x) = x^4 - 2ax^2 - 8bx + a^2$$

$$y^2 = x^3 + ax + b\,.$$

Eliminating x between these equations yields

$$\boxed{(3x^2 + 4a)\varphi_2(x) - (3x^3 - 5ax - 27b)y^2 = \Delta_0\,.}$$

Using $\varphi_2(x) = 4y^2 x_2$ yields the divisibility relation of the theorem.

In most cases, Theorem 1.3 shows that a torsion point (x, y) satisfies the conditions of Theorem 1.4, which then yields an effective and sharp bound for the coordinates of P. If the torsion point is not integral, then Theorem 1.3 still gives bounds for its denominators, and Theorem 1.4 gives corresponding divisibility properties for the coordinates.

Better divisibilities are also given in Cassels [Ca 2], as follows.

Theorem 1.5. *Assume again that π divides the prime number p, and that e is the ramification index. Let $P = (x, y)$ be a point in A_K of order p^s exactly.*
(i) *If $p = 2$, then x, y are integral.*
(ii) *Let $p \neq 2$. Then $P \in A(\pi^r)$ where*

$$r \leqslant e/\phi(p^s),$$

and ϕ is the Euler function. In particular, if $e < p - 1$, then there is no point of order p in $A(\pi)$.

Proof. Let $nP = (x_n, y_n)$ for $1 \leqslant n \leqslant p^s$ and $(n, p) = 1$. There are $\phi(p^s)$ such multiples of P, and they all have exact period p^s. The coordinates x_m are roots of the polynomial

$$g(X) = \psi_{p^s}^2(X)/\psi_{p^{s-1}}^2(X),$$

which has leading coefficient p^2 and integral coefficients (i.e. in $\mathfrak{o}$) by Theorem 2.2 of the preceding chapter. For any polynomial

$$f(X) = a_m X^m + \cdots + a_0 \quad \text{with } a_i \in K,$$

define

$$H(f) = \max |a_i|,$$

where the absolute value is normalized so that $|p| = 1/p$. Then by Gauss' lemma, we have

$$H(f_1 f_2) = H(f_1)H(f_2),$$

and for any $a \in K$, $H(X - a) \geqslant 1$. We have

$$g(X) = p^2 \prod (X - x_n) \prod_Q (X - x(Q)),$$

where the product over Q is taken over the points of exact period p^s other than nP. Furthermore, we have $|x_n| = |x_1|$ for all n with $(n, p) = 1$ by Theorem 1.2. Hence

$$|p^2||x_1|^{\phi(p^s)} \leqslant 1.$$

Since $|p^2| = 1/p^2$ we find

$$|x_1| \leqslant p^{2/\phi(p^s)}.$$

Since $|\pi^e| = |p|$ by definition, and $|x_1| \geqslant |\pi^{-2r}| = p^{2r/e}$, we obtain

$$2r/e \leqslant 2/\phi(p^s),$$

whence $r \leqslant e/\phi(p^s)$ as desired.

Remark. The proof depends on slightly messy arguments relating to the division polynomials. A more conceptual proof will be given in Theorem 3.7.

Let us now specialize K further, and assume that K is a finite extension of the ordinary p-adic field $\mathbf{Q}_p$. Then the ring of integral elements R in K is compact. The set of points A_K may be given a topology in a natural way: Two points are close together if their coordinates are close together, unless one of them is at infinity, i.e. is the origin. In that case, a point is close to the origin if its coordinates (x, y) have a denominator highly divisible by p. This is equivalent to saying that its (t, s) coordinates are themselves highly divisible by p. It is then clear that A_K is a compact topological group. The subgroups $A(\pi^r)$ form open compact subgroups, and in particular are of finite index in A_K.

Theorem 1.6. *If r is sufficiently large, then $A(\pi^r)$ does not contain any torsion point other than the origin. The subgroup of torsion points in A_K is finite.*

Proof. The second assertion clearly follows from the first and the fact that $A_K/A(\pi^r)$ is finite. Let us prove the first.

Let $Q \in A(\pi^r)$ but $Q \notin A(\pi^{r+1})$. Assume that A is given by the usual equation

$$y^2 = x^3 + ax + b , \qquad a, b \in R .$$

The isomorphism $A(\pi^r)/A(\pi^{5r}) \to \pi^r R/\pi^{5r} R$ shows that pQ lies in $A(p\pi^r) = A(\pi^{r+e})$ but not in $A(p\pi^{r+1})$. If $r + e < 5r$, so if $r > e/4$, then $pQ \neq O$. Hence $A(\pi^r)$ does not contain any element of period p. It does not contain any element of period prime to p either by Corollary 1 of Theorem 1.1. This concludes the proof.

This theorem will be refined in § 3 when we consider the log.

§ 2. The Lutz–Nagell Theorem

Let K be an algebraic number field, and $\mathfrak{o} = \mathfrak{o}_K$ the ring of algebraic integers in K. For each prime ideal $\mathfrak{p}$ of $\mathfrak{o}$ we can form the local ring $\mathfrak{o}_\mathfrak{p}$ consisting of all quotients α/β, with $\alpha, \beta \in \mathfrak{o}$ and $\beta \notin \mathfrak{p}$. It is an elementary fact of algebraic number theory that $\mathfrak{o}_\mathfrak{p}$ is a ring of the type considered in the preceding section, in other words, it is a unique factorization domain with exactly one prime element (up to units). Furthermore, one has

$$\mathfrak{o} = \bigcap_{\mathfrak{p}} \mathfrak{o}_\mathfrak{p} ,$$

where the intersection is taken over all prime ideals of $\mathfrak{o}$.

We are interested in the case when the elliptic curve A defined by the equation

$$y^2 = x^3 + ax + b$$

has coefficients $a, b \in \mathfrak{o}$, in other words, integral coefficients, and we are interested in its integral points, i.e. points such that $x, y \in \mathfrak{o}$. In the light of the above remarks, we see that $x \in \mathfrak{o}$ if and only if $x \in \mathfrak{o}_\mathfrak{p}$ for all prime ideals $\mathfrak{p}$, and the results of the preceding section can be applied to each ring $\mathfrak{o}_\mathfrak{p}$. In particular, if we apply Theorem 1.3 concerning torsion points to the present case, we find:

Theorem 2.1. *Let A be defined over a number field K by*

$$y^2 = x^3 + ax + b, \qquad a, b \in \mathfrak{o}_K.$$

Let $P \in A(K)$ have exact period n.
(i) *If n is not a prime power then $x(P) \in \mathfrak{o}_K$.*
(ii) *If $n = p^m$ is a prime power, then*

$$\text{den } x(P) \textit{ divides } \prod_{\mathfrak{p} | p} \mathfrak{p}^{r(\mathfrak{p})}$$

where $r(\mathfrak{p})$ are integers such that $0 \leqslant r(\mathfrak{p}) \leqslant e(\mathfrak{p})/4$, and $e(\mathfrak{p})$ is the ramification index.

Specializing still further to the rational numbers, and using Theorem 1.4, we get:

Theorem 2.2. (Lutz–Nagell) *Let A be defined over the rational numbers by the equation*

$$y^2 = x^3 + ax + b, \qquad a, b \in \mathbf{Z}.$$

Then any torsion point of $A(\mathbf{Q})$ has coordinates (x, y) in $\mathbf{Z}$. If $y \neq 0$ then $y^2 \mid \Delta_0$.

Remark. The Lutz–Nagell theorem holds as well when the equation is

$$y^2 = x^3 + ax^2 + bx + c,$$

and $a, b, c \in \mathbf{Z}$. The only difference is a change from 4 to 2 in Theorem 2.1, in the light of the remarks at the end of the preceding section.

§ 3. The Formal Group

For this section, we shall treat the situation quite generally, and everything we say applies to the most general elliptic curve in any characteristic, defined by an equation

$$y^2 + a_1xy + a_3y = x^3 + a_2x^2 + a_4x + a_6,$$

assuming that the coefficients a_i lie in the ring of integers $\mathfrak{o}$ of a field K *complete under a non-archimedean valuation*, such that the prime number p lies in the maximal ideal $\mathfrak{p}$ of $\mathfrak{o}$. If $x \in K$ we let

$$\operatorname{ord} x = \operatorname{ord}_p x = r \text{ if and only if } |x| = 1/p^r .$$

The reader may however continue to think of the elliptic curve as defined by the simple Weierstrass equation, and may assume that K has characteristic 0. If he looks at Tate [Ta 1], he will see that the proofs go through in general.

We let as before

$$t = x/y .$$

Let r be a positive real number. We now use the notation:

$A(r)$ = set of points P in A_K such that P is not integral, and

$$\operatorname{ord} t(P) \geqslant r .$$

We include the origin in $A(r)$ by convention, and then $A(r)$ is a subgroup of A_K, by Theorem 1.1.

Theorem 3.1. *Let $s = 1/y$. Then there is a power series expansion*

$$s = t^3(1 + A_1 t + A_2 t^2 + \cdots)$$

where A_n is a polynomial of weight n in the a_i with coefficients $\geqslant 0$ in $\mathbf{Z}$.

Proof. We deal with the simple Weierstrass form

$$s = t^3 + as^2 t + bs^3 .$$

Suppose inductively we have found

$$s_m = t^3(1 + A_1 t + \cdots + A_m t^m)$$

such that

$$s_m \equiv t^3 + as_m^2 t + bs_m^3 \pmod{t^{m+4}} .$$

We want to find s_{m+1} to satisfy the congruence mod t^{m+5}. To make the coefficient of t^{m+5} on the left equal to the coefficient of t^{m+5} on the right, we see at once that it suffices that A_{m+1} be a polynomial in a, b and s_k (with $k \leqslant m$) with positive integer coefficients, as desired.

For instance, we have $s = t^3(1 + at^7 + \text{higher terms})$.

Let us denote by $\mathfrak{p}(r)$ or $\mathfrak{p}_r$ the set of elements $z \in \mathfrak{p}$ such that

$$\operatorname{ord} z \geqslant r .$$

Theorem 3.2. *The map*

$$P \mapsto t(P)$$

establishes a bijection between $A(r)$ and $\mathfrak{p}(r)$.

Proof. Given a value t in $\mathfrak{p}(r)$, we solve for s as in the previous theorem, and then for x, y to see that the map is surjective. The same theorem gives the injectivity.

By a **formal group over** $\mathfrak{o}$ we mean a power series $F(X, Y)$ in two variables, with coefficients in $\mathfrak{o}$, such that

$$F(0, Y) = Y, \qquad F(X, 0) = X$$

and

$$F(Z, F(X, Y)) = F(F(Z, X), Y) .$$

Addition on the elliptic curve gives rise to a formal group.

Theorem 3.3. *There exists a formal group*

$$F(X, Y) \in \mathbf{Z}[a, b][[X, Y]]$$

such that if $t_i = t(P_i)$ $(i = 1, 2, 3)$ and $P_1 + P_2 = P_3$ are points in $A(r)$, then

$$t_3 = F(t_1, t_2) .$$

Proof. We go back to formula (1) in § 1 which gives the slope. The denominator can be got rid of by using the geometric series, and similarly in (2). The expression for t_3 as a power series is then immediate.

In particular we see that the points of $\mathfrak{p}$ form a group under the group law whose composition is given by F. We denote this group by $\mathscr{A}(\mathfrak{p})$, and we denote by $\mathscr{A}(\mathfrak{p}_r)$ or $\mathscr{A}(r)$ the subgroup of elements in $\mathfrak{p}_r$, under this law of composition.

The formal group $F(X, Y)$ always starts with a linear term

$$F(X, Y) = X + Y + \text{terms of degree} \geqslant 2 .$$

For each integer $n \geqslant 1$ we have, formally

$$t(nP) = \Psi_n(t(P)) ,$$

where the series Ψ_n are defined inductively by

$$\Psi_1(T) = T, \quad \text{and} \quad \Psi_{n+1}(T) = F(T, \Psi_n(T)).$$

From the linearity of F (mod order 2) we then see that

$$\Psi_n(T) = nT + \text{higher order terms}.$$

It is customary to use the notation $[n]$ to denote the action of an integer n on the formal group. Thus we write

$$[n]T = \Psi_n(T),$$

and

$$[n]t = \Psi_n(t).$$

Under our assumption that the prime number p lies in the maximal ideal $\mathfrak{p}$, it follows that the action of $\mathbf{Z}$ is p-adically continuous, i.e. that for $t \in \mathfrak{p}$,

$$[p^k]t \to 0 \quad \text{as} \quad k \to \infty.$$

This is an easy recursion, left to the reader. By continuity, we may therefore let the p-adic integers $\mathbf{Z}_p$ act on $\mathscr{A}(\mathfrak{p})$, and we use again $[\alpha]$ for the action of an element $\alpha \in \mathbf{Z}_p$ on $\mathscr{A}(\mathfrak{p})$.

From theorem 3.1 we conclude at once that we can express x, y as power series

$$x = t^{-2} + b_1 t^{-1} - b_2 + b_3 t + \cdots$$

$$y = t^{-3} + c_1 t^{-2} + \cdots$$

whose coefficients lie in $\mathbf{Z}[a_1, \ldots, a_6]$. We note that these will essentially be the same series as those coming from analysis. For some purposes, instead of $t = x/y$ it is better to take

$$\boxed{z = -t = -x/y}$$

as the local parameter, so that the power series for y begins with $-z^{-3}$. Such a normalization with z instead of t fits better in some applications, for instance in the next theorem. It is the one selected by Tate [Ta 1].

We let

$$\omega = \frac{dx}{2y + a_1 x + a_3} \left(= \frac{d\wp(z)}{\wp'(z)} = dz \right)$$

be the differential of first kind.

Theorem 3.4. *One can write* $\omega(z) = H(z)\,dz$, *where*

$$H(z) = 1 + h_1 z + h_2 z^2 + \cdots,$$

and $h_n \in \mathbf{Z}[a_1, \ldots, a_6]$.

Proof. In Weierstrass form,

$$\begin{aligned} \omega/dz &= \frac{dx/dz}{2y} = \frac{-2z^{-3} + \cdots}{-2z^{-3} + \cdots} \\ &= \frac{dy/dz}{3x^2 + a} = \frac{-3z^{-4} + \cdots}{-3z^{-4} + \cdots} \end{aligned}$$

has coefficients in $\mathbf{Z}[\frac{1}{2}, a_1, \ldots, a_6]$ but also in $\mathbf{Z}[\frac{1}{3}, a_1, \ldots, a_6]$.

We define the **elliptic logarithm**

$$L(z) = \int H(z)\,dz = z + h_1 \frac{z^2}{2} + h_2 \frac{z^3}{3} + \cdots.$$

Theorem 3.5. *The map* $z \mapsto L(z)$ *is a homomorphism of* $\mathscr{A}(\mathfrak{p})$ *into* K. *If* $r > 1/(p - 1)$ *then*

$$L: z \mapsto L(z)$$

is an isomorphism

$$\mathscr{A}(r) \to \mathfrak{p}(r), \textit{ with ordinary addition.}$$

Proof. In most books on algebraic number theory or class field theory (e.g. mine, Chapter IX, § 2) the reader will find a proof that the log series converges. Although the proof is given for the "ordinary" log, it is valid in general, because it compares the orders of z^n and n at the p-adic valuation.

The fact that L is a homomorphism amounts to a formal identity in the coefficients of the power series involved, as polynomials in $\mathbf{Q}[a_1, \ldots, a_6]$, or $\mathbf{Q}[a, b]$ in Weierstrass form. Since the formula showing that L is a homomorphism holds when a, b are complex numbers, and z_1, z_2, z_3 are complex numbers near the origin, it follows that the desired polynomial identities hold formally, and hence hold in the complete field K. Since the formal series converge in K, we conclude that the addition formula also holds in the present case.

Furthermore, if $r > 1/(p - 1)$ and $\operatorname{ord} z \geqslant r$ then the same references on the p-adic logarithm show that

$$\operatorname{ord} z = \operatorname{ord} L(z).$$

Indeed, one can construct an exponential power series which is formally the inverse of the log, and converges absolutely in the disc $\operatorname{ord} z > 1/(p - 1)$. Hence L is injective on $\mathscr{A}(r)$. We leave the surjectivity of $\mathscr{A}(r)$ onto $\mathfrak{p}(r)$ as an exercise to the reader.

Note. We used the crutch of complex analysis to carry us through the preceding proof. Analogous statements are true on all formal groups. Cf. Fröhlich, *Formal Groups*, Springer Lecture Notes 74.

The foundations of formal groups as we are using them now (commutative, one variable) are due to Lubin [Lu], and useful further results are due to Lubin–Tate [Lu–Ta].

We return to some applications of the formal group. We have seen that $\mathbf{Z}_p$ operates on the formal group. Recall that the group μ_{p-1} of $(p-1)$th roots of unity is contained in $\mathbf{Z}_p$. Let ζ be a primitive $(p-1)$th root of unity, i.e. a generator of μ_{p-1}. Then as mentioned previously, $[\zeta]$ denotes the action of ζ on the formal group.

Lubin's Lemma. *There exists a formal power series*

$$T = g(t) \in \mathfrak{o}[[t]]$$

such that $T = t +$ higher terms, and

$$[\zeta]T = \zeta T.$$

If Z is the power series such that $[\zeta]t = Z(t)$, this means that

$$g(Z(t)) = \zeta g(t).$$

Proof. Assume recursively that we have found an approximate solution such that

$$[\zeta]t = \zeta t + \lambda t^n + \cdots$$

with $\lambda \in \mathfrak{o}$. Let $T = t + at^n$. We must find a such that

$$[\zeta]T \equiv \zeta T \bmod T^{n+1}.$$

But

$$\begin{aligned} [\zeta]T &\equiv \zeta t + \lambda t^n + a(\zeta t + \lambda t^n)^n \pmod{T^{n+1}} \\ &\equiv \zeta t + (\lambda + a\zeta^n)t^n \pmod{T^{n+1}}, \end{aligned}$$

and we want this to be

$$\equiv \zeta t + \zeta a t^n \pmod{T^{n+1}}.$$

This amounts to solving

$$\lambda + a\zeta^n = \zeta a, \quad \text{or} \quad a(\zeta - \zeta^n) = \lambda\,.$$

If $n \not\equiv 1 \bmod p - 1$, then $\zeta - \zeta^n$ is a unit and we can solve integrally for a. On the other hand, suppose $n \equiv 1 \bmod p - 1$. We then show that $\lambda = 0$, in which case we may solve for a arbitrarily in $\mathfrak{o}$.

To see that $\lambda = 0$, we use the fact that $\zeta^n = \zeta$, whence

$$[\zeta]^n t = t\,.$$

But

$$\begin{aligned} [\zeta]^2 t = [\zeta]([\zeta]t) &\equiv \zeta(\zeta t + \lambda t^n) + \lambda(\zeta t + \lambda t^n)^n \bmod t^{n+1} \\ &\equiv \zeta^2 t + (\zeta\lambda + \lambda\zeta^n)t^n \\ &\equiv \zeta^2 t + 2\zeta\lambda t^n\,. \end{aligned}$$

Inductively,

$$[\zeta]^n t \equiv \zeta^n t + n\zeta^{n-1}\lambda t^n\,.$$

Thus $n\zeta^{n-1}\lambda = 0$, whence $\lambda = 0$, as desired.

We may therefore proceed inductively to define the power series $g(t)$ to conclude the proof of the theorem.

Theorem 3.6. *If T is the parameter such that $[\zeta]T = \zeta T$, and*

$$[p]T = pT + \sum_{n \geqslant 2} c_n T^n\,,$$

then $c_n = 0$ unless $n \equiv 1 \bmod p - 1$. In particular,

$$[p]T = pT + \alpha T^p + \textit{higher terms}\,,$$

for some $\alpha \in \mathfrak{o}$.

Proof. Since $[p]$ and $[\zeta]$ commute, we find that the relations

$$[p][\zeta]T = [p](\zeta T) = p\zeta T + \sum c_n \zeta^n T^n$$
$$[\zeta][p]T = [\zeta](pT + \sum c_n T^n) = \zeta p T + \sum \zeta c_n T^n$$

imply that $a_n = 0$ unless $n \equiv 1 \bmod p - 1$, as desired.

I am indebted to Katz for pointing out that the above results of Lubin can be used to recover Cassels' divisibility properties, as follows.

Theorem 3.7. *Let $t_1 \in \mathfrak{p}$ and $t_1 \neq 0$ be such that $[p]t_1 = 0$, so t_1 is the parameter of a point of order p. Let*

$$[p]t_{k+1} = t_k .$$

Then

$$\operatorname{ord} t_k \leqslant 1/\phi(p^k) .$$

Proof. Since ord t = ord T, we shall work with the parameter T in the Lubin normalization. For $k = 1$ we know that

$$\mathscr{A}(r) \approx \mathfrak{p}(r) \quad \text{for} \quad r > 1/(p-1)$$

under the logarithm map, so there is no torsion in $\mathscr{A}(r)$. This proves the theorem for $k = 1$. Inductively, assume it for k. By the preceding theorem, letting $T = T_{k+1}$, we find

$$T_k = pT + \alpha T^p + \text{higher terms} ,$$

with integral α. Note that ord $T \leqslant 1/(p-1)$. Then

$$\frac{1}{\phi(p^k)} \geqslant \operatorname{ord} T_k \geqslant \min(\operatorname{ord} T + 1, p \operatorname{ord} T)$$

$$\geqslant p \operatorname{ord} T ,$$

whence

$$\operatorname{ord} T \leqslant \frac{1}{p\phi(p^k)} = \frac{1}{\phi(p^{k+1})},$$

as was to be shown.

§ 4. The Néron Function

We again assume that R has a unique prime π, giving rise to a discrete absolute value v. *We assume that K is complete.* We write $v(t) = -\log |t|$ for $t \in K^*$.

Theorem 4.1. *Let A be an elliptic curve defined over K. There exists a unique function λ defined on $A_K - \{O\}$, with real values, satisfying the following conditions:*

(i) *λ is continuous and bounded outside every neighborhood of O.*

(ii) *For one (and therefore every) local parameter z at O, the limit*

$$\lim_{P \to O} \lambda(P) - v(z(P))$$

exists.

(iii) *For all $P, Q \in A_K$ such that $P, Q, P \pm Q \neq O$ we have*

$$\lambda(P+Q) + \lambda(P-Q) = 2\lambda(P) + 2\lambda(Q) + v(x(P) - x(Q)) - \tfrac{1}{6}v(\Delta) .$$

The function of Theorem 4.1 will be called the **Néron function**. Its existence will be proved in the next two sections, distinguishing cases. Uniqueness is obvious, just as in the archimedean case, Theorem 7.2 of Chapter II. The particular formulation of Theorem 4.1 given here is due to Tate.

If E is a finite extension of K, then the absolute value extends uniquely to E, and by uniqueness, the function λ_K is the restriction of λ_E to $A_K - \{O\}$. Thus we were justified in omitting the subscript indexing the field from the notation.

Of course, we should also use an index v, i.e. write λ_v, if we wish to make the reference to v explicit.

Let

$$f \colon A \to B$$

be an isomorphism of elliptic curves defined over some finite extension of K. Uniqueness of the Néron function shows that

$$\lambda_B \circ f = \lambda_A .$$

In fact, suppose that B is given in Weierstrass form

$$y^2 = x^3 + ax + b ,$$

and we make an isomorphism using multiplication by c, so that

$$x' = c^2 x, \; y' = c^3 y, \; \ldots, \; \Delta' = c^{12}\Delta .$$

The expression

$$v(x(P) - x(Q)) - \tfrac{1}{6}v(\Delta)$$

is invariant under this isomorphism, because $v(c^2)$ cancels. Thus to find the Néron function, it suffices to do so on a convenient model in an isomorphism class.

We have seen in § 1 that the set of points (x, y) such that x is not integral forms a group. On this group, the function λ has a simple expression. Up to a normalizing constant, it is merely the filtration itself. Thus for any point $P \in A_K$ with coordinates (x, y) such that x is not integral at v, we define

$$\lambda_0(P) = -\tfrac{1}{2}v(x(P)) .$$

If the absolute value is normalized so that

$$|\pi| = 1/e\,,$$

and if $x = u/\pi^{2r}$ with some unit u, then

$$\lambda_0(P) = r\,.$$

Note that $z = x/y$ is a local parameter at the origin, and therefore that

$$\boxed{\lambda_0(P) = v(z(P))\,.}$$

This function λ_0 satisfies the quadratic property of the Néron function, normalized without the appearance of Δ. More precisely:

Theorem 4.2. *Assume that A is given in Weierstrass form*

$$y^2 = x^3 + ax + b$$

with integral a, b. Let $P, Q \in A(\pi)$, and assume $P, Q \neq O$ and also $P \neq \pm Q$. Then

$$\lambda_0(P+Q) + \lambda_0(P-Q) = 2\lambda_0(P) + 2\lambda_0(Q) + v(x(P) - x(Q))\,.$$

Proof. Let

$$x_1 = x(P), \qquad x_2 = x(Q), \qquad x_3 = x(P+Q), \qquad x_4 = x(P-Q)\,.$$

Here as in the rest of the study of quadraticity, we use the **parallelogram formula**:

PAR. $$x_3 x_4 = \frac{(x_1 x_2 - a)^2 - 4b(x_1 + x_2)}{(x_1 - x_2)^2}$$

left as a routine exercise to the reader.

The formula to be proved amounts to

$$\operatorname{ord} x_3 x_4 = \operatorname{ord} (x_1 x_2)^2/(x_1 - x_2)^2\,.$$

Since x_1, x_2 have a pole at π, this relation is obvious from **PAR**.

Let

$$\bar{R} = R/\mathfrak{p}$$

be the residue class field, and for $c \in R$ we let $\bar{c}$ be its residue class mod $\mathfrak{p}$. The equation

$$y^2 = x^3 + \bar{a}x + \bar{b}$$

defines an elliptic curve, provided that the cubic polynomial on the right has distinct roots (equivalent to non-vanishing discriminant), and the characteristic of $\overline{R}$ is $\neq 2$, 3. If this is the case, we say that A has **good reduction** mod $\mathfrak{p}$ or **non-degenerate reduction** mod $\mathfrak{p}$.

We define the **local height** by the formula

$$h_v(P) = \log \max \{1, |x(P)|_v\} .$$

Theorem 4.3. *Assume that A has non-degenerate reduction* mod $\mathfrak{p}$. *Define*

$$\lambda(P) = \tfrac{1}{2}h_v(P) ,$$

that is:

$$\begin{aligned} \lambda(P) &= -\tfrac{1}{2}v(x(P)) \quad && \textit{if } P \in A(\pi) \\ &= 0 && \textit{otherwise} . \end{aligned}$$

Then λ is the Néron function, and it also satisfies the condition of Theorem 4.2.

Proof. The first two properties of the Néron function are obvious. There remains to prove the property of Theorem 4.2. We have to distinguish cases. If $P = (x, y)$ we put $\overline{P} = (\bar{x}, \bar{y})$ if $x \in R$, and otherwise we put $\overline{P} = \overline{O}$. Theorem 4.2 already covers the case when $\overline{P} = \overline{Q} = \overline{O}$. We deal with the remaining ones.

Case 1. $\overline{P} \pm \overline{Q} \neq \overline{O}$, $\overline{P} \neq \overline{O}$ and $\overline{Q} \neq \overline{O}$.
In this case, all terms in the formula are equal to 0, we are done.

Case 2. $\overline{P} = \overline{O}$ but $\overline{Q} \neq \overline{O}$.
In this case, $x(Q)$ is a unit, but $x(P)$ has a pole at π. The formula is again obvious.

Case 3. $\overline{P} + \overline{Q} = \overline{O}$ and $\overline{Q} \neq \overline{O}$.
This is the most interesting case, where we have to use non-degenerate reduction somewhere.

(a) Suppose first that

$$\overline{P} + \overline{Q} = \overline{O} \quad \text{but } \overline{P} \neq \overline{Q} .$$

Thus $\overline{P} = -\overline{Q}$. The desired formula amounts to

$$-\tfrac{1}{2} \operatorname{ord} x(P + Q) = \operatorname{ord} (x(P) - x(Q)) .$$

But

$$x_3 = -x_1 - x_2 + \left(\frac{y_2 - y_1}{x_2 - x_1}\right)^2.$$

By hypothesis, $\bar{x}_2 = \bar{x}_1$. Then $\bar{y}_2 \neq \bar{y}_1$, otherwise $\bar{P} = \bar{Q}$. Hence the formula is clear because $y_2 - y_1$ is a unit.

(b) $$\bar{P} + \bar{Q} = \bar{O} \quad \text{and} \quad \bar{P} = \bar{Q} \quad \text{so} \quad 2\bar{P} = \bar{O}.$$

Then the formula amounts to

$$-\tfrac{1}{2} \operatorname{ord} x_3 x_4 = \operatorname{ord}(x_1 - x_2)$$

But the numerator of the right-hand side of **PAR** with x_1, x_2 replaced by $\bar{x}$, is

$$(\bar{x} - \bar{a})^2 - 4\bar{b}\bar{x} = -8\bar{x}\bar{y}^2 + (3\bar{x}^2 + \bar{a})^2 = (3\bar{x} + \bar{a})^2 \neq 0,$$

because $\bar{x}$ is a root of

$$f(\bar{x}) = \bar{x}^3 + \bar{a}\bar{x} + \bar{b} = 0,$$

and is not a multiple root by good reduction. Hence $3x^2 + a$ is a unit. Hence the numerator on the right hand side of **PAR** is a unit, and the desired order relation is now obvious.

Remark. When the characteristic of the residue class field is 2 or 3, the ordinary Weierstrass form with which we have dealt is of course insufficient. One has to deal with other normal forms due to Deuring, cf. [L 2], Appendix 1 and Tate [Ta 1]. Especially, the equation for the curve has to be taken in the form

$$y^2 + a_1 xy + a_3 y = x^3 + a_2 x^2 + a_4 x + a_6.$$

The indices on a_j indicates the "weight". It is then true that Theorems 4.1, 4.2, and 4.3 remain valid. The proof is the same, mutatis mutandis. All coefficients $a_1, \ldots, a_6$ are again assumed to be integral, and the local height is again defined by the same formula in terms of the function x.

We wish to investigate how the Néron function differs from a given model for the elliptic curve to another model. We work out first the simplest case.

Theorem 4.4. *Assume that A is given in the simple Weierstrass form*

$$y^2 = x^3 + ax + b,$$

with integral a, b, and that the characteristic of the residue class field is not 2 or 3.

Assume $v(j) \geqslant 0$. *Let* $c^{12} = 1/\Delta$. *Then*

$$\lambda_v(P) = \tfrac{1}{2} \log \max \{1, |c^2 x(P)|\}$$

whence

$$0 \leqslant \lambda_v - \tfrac{1}{2} h_v \leqslant \tfrac{1}{12} v(\Delta) .$$

Proof. In this case, $\Delta = -16(4a^3 + 27b^2)$. If we let

$$a' = c^4 a, \qquad b' = c^6 b, \qquad \Delta' = c^{12}\Delta ,$$

then the elliptic curve A' defined by

$$y'^2 = x'^3 + a'x + b'$$

has discriminant Δ', which is a unit, and A' is isomorphic to A under the correspondence

$$x' = c^2 x, \qquad y' = c^3 y .$$

We have $j = -(48a)^3/\Delta$. The hypothesis $v(j) \geqslant 0$ immediately implies that a', b' are integral, and the curve A' has good reduction (at the absolute value extending v in the field $K(c)$). Hence if $P = (x, y)$ corresponds to $P' = (x', y')$ under the isomorphism, we have

$$\lambda_A(P) = \lambda_{A'}(P') = \tfrac{1}{2} \log \max \{1, |c^2 x|\} ,$$

from which the theorem is obvious.

Theorem 4.5. (Tate) *Let A be an elliptic curve defined over the complete field K. Let Δ be its discriminant. Assume that A is defined by an equation*

$$y^2 + a_1 xy + a_3 y = x^3 + a_2 x^2 + a_4 x + a_6 ,$$

and that the coefficients are v-integral. We always have the inequalities

$$\min \{0, \tfrac{1}{24} v(j)\} \leqslant \lambda_v - \tfrac{1}{2} h_v \leqslant \tfrac{1}{12} v(\Delta) .$$

Proof. The proof is separated into cases.

Case 1. $v(j) \geqslant 0$. Thus j is integral. In this case, the curve is isomorphic to a curve with non-degenerate reduction. We have dealt with the situation when the characteristic of the residue class field is not 2 or 3 in Theorem 4.4. When the characteristic is 2 or 3, then one cannot use the simplest Weierstrass form, but one must use the other canonical forms of Deuring, as tabulated by Tate, cf. [Ta 1] and [L 2], Appendix 1. The isomorphism is given in a more complicated way than before, but the explicit formulas are manageable, and one finds again the estimate with $\frac{1}{12}v(\Delta)$ on the right hand side. In the present case, the left hand side is 0. The reader can slug it out if he wishes.

Case 2. $v(j) < 0$. Thus j is not integral, $|j| > 1$. This is the case for which one has a good analytic model for a curve isomorphic to A, the Tate curve, and this case will be discussed in the next section.

Remark. Estimates for the difference of the Weil height and the Néron–Tate height have been given by Demjanenko [De 7] and Zimmer [Z], see also Colliot–Thélène [C–T] for a special case.

§ 5. The Tate Curve

In this section, due to Tate, we assume that the reader is acquainted with the Tate curve over a field K complete under a non-archimedean absolute value v. All we need of that theory is given in [L 2], Chapter 15. We recall merely that the Tate curve is associated with a non-integral j-invariant, corresponding to an element q of K with $|q| < 1$.

We let $t \in K^*$ be a multiplicative variable, and for $t \neq q^n$ for all integers n we define

$$g_0(t) = (t-1)\prod_{n=1}^{\infty}(1-q^n t)(1-q^n/t)\,.$$

Then the product converges absolutely, and it is directly verified that g_0 satisfies the functional equation

$$\boxed{g_0(qt) = g_0(t^{-1}) = -\frac{1}{t}\,g_0(t)\,.}$$

We let $v(t) = -\log|t|$, and

$$\alpha(t) = \frac{v(t)}{v(q)}\cdot$$

We define the function λ in this case to be

$$\boxed{\lambda(t) = v(g_0(t)) + \tfrac{1}{2}B_2(\alpha(t))v(q)\,.}$$

Lemma 1. *The function λ depends only on t* mod $q^{\mathbf{Z}}$, i.e.

$$\lambda(qt) = \lambda(t).$$

Proof. From the functional equation we know that

$$v(g_0(qt)) = -v(t) + v(g_0(t)) .$$

Since

$$\alpha(qt) = 1 + \alpha(t) ,$$

the lemma is then obvious by writing down how the Bernoulli polynomial $B_2(\alpha)$ changes under $\alpha \mapsto \alpha + 1$.

From the lemma, we see that λ is defined on $K^*/q^{\mathbf{Z}}$ (except on $q^{\mathbf{Z}}$ itself), and therefore λ is defined on the Tate curve which is analytically isomorphic to $K^*/q^{\mathbf{Z}}$ (except at the origin). From [L 2], Chapter 15, we know that the x-coordinate of the Tate curve is parametrized by

$$X(t) = \frac{t}{(1-t)^2} + \sum_{n=1}^{\infty} \left(\frac{q^n t}{(1-q^n t)^2} + \frac{q^n t^{-1}}{(1-q^n/t)^2} - \frac{q^n}{(1-q^n)^2} \right). \tag{1}$$

We have of course periodicity, $X(qt) = X(t)$. In the restricted range

$$|q| < |t| < |q^{-1}|$$

we also have the expression

$$X(t) = \frac{t}{(1-t)^2} + \sum_{n=1}^{\infty} \frac{nq^n}{1-q^n} (t^n + t^{-n} - 2) . \tag{2}$$

Lemma 2. *$X(t)$ has a pole at the valuation if and only if $t \equiv 1 \pmod{\mathfrak{p}}$, where $\mathfrak{p}$ is the maximal ideal, and in that case,*

$$v(X(t)) = -2v(1-t) .$$

Proof. Obvious from the series for X.

Since λ depends only on the class of t mod $q^{\mathbf{Z}}$, we may view λ as a function of points on the Tate curve, and therefore we write

$$\lambda(P) = \lambda(P_t)$$

if $X(t)$ is the X-coordinate of P. By periodicity, there exists a unique t such that $P = P_t$ and

$$0 \leqslant \alpha(t) < 1, \quad \text{or equivalently,} \quad |q| < |t| \leqslant 1 .$$

We shall write $t = t_P$ for this unique t, and $\alpha(P) = \alpha(t_P)$.

Theorem 5.1. *For the Tate curve, we have the expression*

(i) $$\lambda(P) = v(1 - t_P) + \tfrac{1}{2}B_2(\alpha(P))v(q)\,.$$

Also,

(ii) $$\lambda(P) = \begin{cases} \tfrac{1}{2}B_2(\alpha(P))v(q) & \text{if } X(P) \text{ is integral} \\ v(z(P)) + \tfrac{1}{12}v(q) & \text{if } X(P) \text{ is not integral,} \end{cases}$$

and z is a local parameter at the origin such that

$$v(z(P)) = -\tfrac{1}{2}v(X(P))$$

for all non-integral points P (for instance, $z = X/Y$).

Proof. For t normalized so that $|q| < |t| \leqslant 1$, the infinite product for $g_0(t)$ is a unit, and the only contribution to λ from g_0 comes from the factor $(t - 1)$. This makes (i) obvious. As for (ii), we know that if X is integral, then $v(1 - t_P) = 0$ from Lemma 2. On the other hand, if X is not integral, then Lemma 2 gives the exact order, and we know that X has a pole of order equal to twice the order of a local parameter at the origin. In this case, $v(t_P) = 0$, $\alpha(P) = 0$, and hence

$$\lambda(P) = v(z(P)) + \tfrac{1}{12}v(q)$$

as was to be shown.

We have already seen that the first two properties of a Néron function are satisfied.

Theorem 5.2. *The function λ as defined above is the Néron function on the Tate curve.*

Proof. There remains only to check the third property. The formula from Theorem 8.2, Chapter II is valid in the present case, when $t \in K^*$ and $t \neq q^n$ for any integer n. In other words, putting

$$\psi(t) = t^{-\frac{1}{2}}(t - 1) \prod_{n=1}^{\infty} (1 - q^n t)(1 - q^n/t)$$

$$\Delta = q \prod_{n=1}^{\infty} (1 - q^n)^{24}$$

we have

$$X(t) - X(t') = -\frac{\psi(tt')\psi(t/t')}{\psi(t)^2\psi(t')^2}(\Delta/q)^{\frac{1}{6}}$$

for t, t', tt', $t/t' \neq q^n$ for all integers n. We note that $\lambda(t)$ can be decomposed into the sum

$$\begin{aligned}\lambda(t) &= v(g_0(t)) + \tfrac{1}{2}\alpha^2 v(q) - \tfrac{1}{2}\alpha v(q) + \tfrac{1}{12}v(q)\\ &= v(g_0(t)) + \tfrac{1}{2}\alpha^2 v(q) - \tfrac{1}{2}v(t) + \tfrac{1}{12}v(q)\end{aligned}$$

so

$$\boxed{\lambda(t) = v(\psi(t)) + \tfrac{1}{12}v(q) + \tfrac{1}{2}\alpha^2 v(q)\,.}$$

The term involving $\alpha^2 = \alpha(t)^2$ will cancel in the expression

$$\lambda(tt') + \lambda(t/t') - 2\lambda(t) - 2\lambda(t')\,.$$

The extra term $\frac{1}{12}v(q)$ gives rise to $-\frac{1}{6}v(q)$ in the quadraticity formula for the Néron function. Hence the property is obvious.

The possibility of constructing other related local "heights" has been considered by Manin [Man 1].

On the Tate curve we may now prove the estimate giving the difference between the Néron function and the local height

$$h_v(P) = \log \max\{1, |X(P)|_v\}\,.$$

Corollary. *On the Tate curve, we have the estimate*

$$-\tfrac{1}{24}v(q) \leqslant \lambda(P) - \tfrac{1}{2}h_v(P) \leqslant \tfrac{1}{12}v(q)\,.$$

Proof. This comes from looking at the two cases when $X(P)$ is integral or not integral, and noting that $B_2(\alpha)$ has minimum $-\frac{1}{12}$ and maximum $\frac{1}{6}$ on the interval $[0, 1]$.

Suppose we deal with an arbitrary elliptic curve A with $v(j) < 0$, and for simplicity assume it is in ordinary Weierstrass form

$$y^2 = x^3 + ax + b\,.$$

Let Δ be its discriminant. From the theory of the Tate curve, [L 2], Chapter 15, § 1 we know that A is isomorphic to the Tate curve over a quadratic extension at most. Let T be this Tate curve. Then there exists c such that

$$c^{12}\Delta_T = \Delta_A\,,$$

and we write

$$\Delta_A = \Delta \quad \text{and} \quad \Delta_T = \Delta'\,.$$

If X is the coordinate on the Tate curve as before, we put

$$x' = X + \tfrac{1}{12}$$

so that

$$c^2 x' = x \quad \text{and} \quad x = c^2(X + \tfrac{1}{12}) .$$

Then

$$\Delta' \sim q \sim 1/j \quad \text{and} \quad c^{12}\Delta' = \Delta .$$

The sign $\sim$ means that the two quantities have the same order at the valuation.

The local height functions on T and A are given by

$$\log \max \{1, |X|_v\} \quad \text{on } T$$

$$\log \max \{1, |x|_v\} = \log \max \{1, |c^2(X + \tfrac{1}{12})|_v\} \quad \text{on } A .$$

The Néron function on A is that composed with the isomorphism between A and T. We have

$$c^{12} \sim \Delta/q \sim \Delta j ,$$

and also

$$j = -\frac{(3 \cdot 2^4)^3 a^3}{\Delta} \cdot$$

In particular,

$$|c^{12}| = |\Delta j| = |(3 \cdot 2^4)^3| |a^3| .$$

and

$$c^2 \sim 3^{\frac{1}{2}} 2^2 a^{\frac{1}{2}} .$$

Therefore, if a is v-integral, it follows that c is also integral. We wish to prove Theorem 4.5. We have:

$$\text{(1)} \qquad \lambda(P) - \tfrac{1}{2}h_v(P) = \lambda_T(P') - \tfrac{1}{2}h_{T,v}(P') + \tfrac{1}{2}[h_{T,v}(P') - h_v(P)]$$

$$\text{(2)} \qquad \leqslant \tfrac{1}{12}v(q) + \tfrac{1}{2}[h_{T,v}(P') - h_v(P)] ,$$

where the point P on A corresponds to the point P' on T. We shall need a lemma.

Lemma. *Assume that a, b are v-integral. Then*

$$0 \leqslant \tfrac{1}{2}[h_{T,v}(P') - h_v(P)] \leqslant \tfrac{1}{12}[v(\Delta) - v(q)] = v(c) .$$

Proof. We distinguish cases.

Case 1. X is integral. If the valuation does not divide 2 or 3, the lemma is obvious. Suppose v divides 3. Since

$$\Delta = -16(4a^3 + 27b^2)\,,$$

and j is not integral, it follows that 27 divides Δ, so 27 divides a^3, therefore 3 divides a. Hence 3 divides c^2, and x is also integral, so both local heights are 0. A similar argument applies if v divides 2, and this case is proved.

Case 2. X is not v-integral. Then

$$h_{T,v}(P') = \log |X|_v, \qquad h_v(P) = \log \max \{1, |c^2(X + \tfrac{1}{12})|\}\,.$$

The estimate follows at once by considering the subcases $1 < |X| \leqslant |\frac{1}{12}|$ and $|X| > |\frac{1}{12}|$. This proves the lemma.

We return to tne proof of the theorem. First using (1) and the positive property of the lemma, we get

$$\begin{aligned}\lambda(P) - \tfrac{1}{2}h_v(P) \geqslant -\tfrac{1}{24}v(q) &= -\tfrac{1}{24}v(\Delta') \\ &= -\tfrac{1}{24}(v(\Delta) - 12v(c)) \\ &= \tfrac{1}{24}v(j)\end{aligned}$$

as follows at once from the expression for j in terms of 2, 3, a, Δ. This proves one of the desired inequalities. The other inequality follows equally easily from the lemma and (2). This concludes the proof of Theorem 4.5, i.e. the description of how much the Néron function differs from the local height in the case of a curve isomorphic to the Tate curve.

§ 6. *p*-Adic Points of Order *p*

The theorem in this section gives a further application of the integrality statements for torsion points.

We let μ_p be the group of p-th roots of unity, p prime $\geqq 5$. We let F be a finite extension of $\mathbf{Q}_p$ and assume that A is defined over F. We let e be the ramification index of F over $\mathbf{Q}_p$.

Theorem 6.1. *Assume that A has a point of order p rational over F. Assume also that $e < p - 1$ if the invariant j_A is p-integral, and $2e < p - 1$ if j_A is not p-integral. Then*

$$F(A_p) = F(\mu_p)\,.$$

Proof. The field $F(A_p)$ is Galois over F. Let G be its Galois group. There is a canonical skew-symmetric pairing on the torsion points of A, cf. for instance [L 2], Chapter 18, § 1 which we denote by $\langle P, Q\rangle$. It is a Galois pairing, in the sense that for any automorphism σ of the algebraic closure of F over F, we have

$$\langle \sigma P, \sigma Q\rangle = \sigma\langle P, Q\rangle .$$

Let P be the point of order p in A_F, and Z the cyclic group generated by P. Then

$$Q \mapsto \langle P, Q\rangle$$

gives a G-isomorphism of A_p/Z with μ_p.

We shall prove the theorem only under a slightly stronger hypothesis, namely that $we < p - 1$, where w is the number of automorphisms of A (over the algebraic closure of $\mathbf{Q}_p$).

Case 1. The invariant j_A is not p-integral. Then A is isomorphic to the Tate curve over an extension E of F, of degree 1 or 2. (For the Tate curve, cf. [L 2], Chapter 15.)

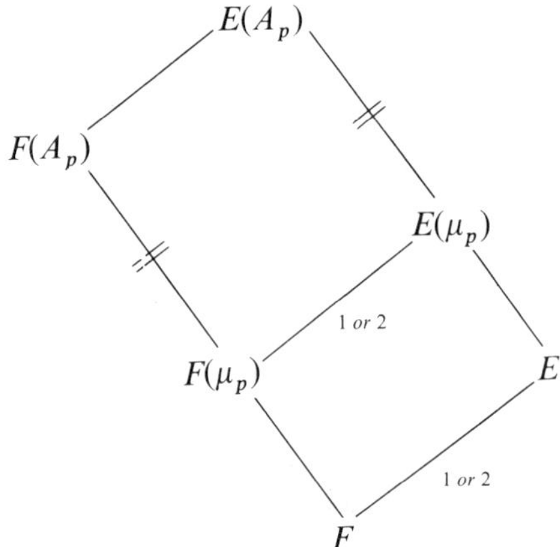

Since $2e < p - 1$, the roots of unity μ_p are not in E. Under the Tate parametrization, we have an isomorphism

$$A_E \approx E^*/\{q\} ,$$

and A_p is isomorphic to

$$\{q^{1/p}, \mu_p\}/\{q\} .$$

Hence the point P corresponds to a choice of $q^{1/p}$ under this isomorphism (i.e. does not correspond to a pth root of unity). Thus $q^{1/p} \in E$, whence

$$E(\mu_p) = E(A_p) ,$$

and from the field diagram, it follows that $F(A_p) = F(\mu_p)$, as desired.

Remark. Tate has pointed out that the condition $2e < p - 1$ is best possible. Indeed, let F be the subfield of $\mathbf{Q}(\mu_p)$ of degree $(p-1)/2$ over $\mathbf{Q}_p$. Let q be a prime element of F, and let $T = T(q)$ be the Tate curve over F. It is easy to construct an elliptic curve A over F which is isomorphic to T over $E = \mathbf{Q}(\mu_p)$ but not over F. (If T is defined by a usual equation

$$y^2 = x^3 + ax + b\,,$$

one easily finds a suitable c such that replacing a by a^4c and b by c^6b yields the desired equation for A.) The group of points A_F then corresponds to those points Q in T_E such that

$$Q + \sigma Q = O\,,$$

where σ is the automorphism of E over F. In the Tate model, this amounts to the multiplicative condition

$$\lambda\lambda^\sigma = q^m$$

for $\lambda \in E^*$ and some integer m which can be taken to be 0 or 1. The p-th roots of unity ζ in μ_p do satisfy this condition, that is $\zeta^\sigma = \zeta^{-1}$, so they give rise to a point of order p in A_F.

Case 2. The invariant j_A is integral at p. We shall first carry out the proof under the hypothesis that A has good reduction at $\mathfrak{p}$, the maximal ideal of the ring of integers $\mathfrak{o}$ in F. By Theorem 1.5 or 3.7 the group Z cannot reduce to O mod $\mathfrak{p}$. (A point reduces to O mod $\mathfrak{p}$ if and only if its x and y coordinates are not integral.) Let a bar denote reduction mod $\mathfrak{p}$, and let $K = F(A_p)$. We have the reduction homomorphism

$$A(F) \to \overline{A}(\overline{F}) \supset \overline{Z} \approx \mathbf{Z}/p\mathbf{Z}\,.$$

Let Z' be the kernel of $A_p \to \overline{A}_p$ in $A(K)$. Then Z' is a subgroup of order p (cf. [L 2], Chapter 13), and Z' is G-isomorphic to μ_p. Hence

$$F(Z') = F(A_p) = F(\mu_p)\,,$$

as desired.

It is true in general that under the existence of a rational point of order $\geqslant 5$ the curve A must have good reduction already over $\mathfrak{o}$, but this takes the theory of moduli to prove.

If one wishes to avoid the use of such deeper technique, then one must use the fact that after an extension E of degree w (depending on the existence of corresponding automorphisms of A) the curve has good reduction over E. This is proved by the same argument as in Theorem 4.4, making an isomorphism

$$a' = c^4 a\,, \qquad b' = c^6 b\,, \qquad \Delta' = c^{12}\Delta\,,$$

so that the elliptic curve A' defined by

$$y'^2 = x'^3 + a'x' + b'$$

has discriminant Δ', which is a unit, and its coefficients a', b' are p-integral under the assumption that a, b are p-integral. Using the field E one can argue as before, except that the ramification index of E over $\mathbf{Q}_p$ may now be slightly bigger, namely we, and hence to use Theorem 1.5 one must assume the stronger condition that $we < p - 1$. Then the same proof as above goes through.

The theorem of this section is of Lutz–Nagell type, but evolved from more modern points of view concerning the classification of group schemes of order p, as in Oort–Tate [OT], Raynaud [Ra], and Fontaine [Fo]. I learned it from the important use made by Mazur [Maz] in his proof of the finiteness of rational points on certain factors of the Jacobians of modular curves.

Chapter IV. Heights

In 1922, Mordell proved a conjecture of Poincaré that the group of rational points on an elliptic curve is finitely generated. The reader can profitably look at the exposition in his book [Mo]. In 1929–1930, Weil extended this to abelian varieties [We 1].

The proof has always, more or less, been separated into two steps. The first shows that A_K/mA_K is finite for some integer m, classically taken to be $m = 2$. The second gives an infinite descent, using heights. The theory of heights has been developed substantially since that time. One of the principal more recent discoveries was that the (logarithmic) height differed by a bounded function from a quadratic form (Néron–Tate, see Néron [Ne 2] and the Bourbaki seminar report [L 9]). In this chapter, we develop the theory of heights and show how it applies to the descent.

In the next chapter, we give the proof that A_K/mA_K is finite, a result usually known as the weak Mordell–Weil theorem.

The exposition of this chapter benefited both from Cassel's survey [Ca 1] and Tate's Phillips Lecture Notes at Haverford. Large chunks of theory required to present the proofs on abelian varieties as in *Diophantine Geometry* are here replaced by computational formulas, basically centering around the addition formula of the Weierstrass function.

§ 1. Basic Properties

Let $\mathbf{P}^N$ denote projective N-space. If K is a field, we let $\mathbf{P}^N(K)$ denote the set of points in $\mathbf{P}^N$ which are rational over K, i.e. such that the points have coordinates

$$P = (a_0, \ldots, a_N)$$

with $a_i \in K$ for all i.

We shall deal with number fields, i.e. finite extensions of the rational numbers. On $\mathbf{Q}$, we have the usual set of absolute values $M_{\mathbf{Q}}$. These are of two types: The ordinary absolute value, also called the absolute value at infinity, denoted by $|\ |$ or $|\ |_\infty$, induced by the ordinary absolute value on the real numbers. The other type is the p-adic absolute value $|\ |_p$ for each prime p. If $a = p^r m/n$, where m, n are integers not divisible by p, then by definition

$$|a|_p = 1/p^r .$$

These absolute values are also denoted by v_∞ or v_p respectively.

Let P be a point in $P^N(\mathbf{Q})$ with coordinates $a_i \in \mathbf{Q}$ as above. We define its **height**

$$H_{\mathbf{Q}}(P) = \prod_{v \in M\mathbf{Q}} \max |a_i|_v .$$

This is independent of the choice of coordinates, because we have the product formula

$$\prod_v |c|_v = 1 , \quad \text{for } c \in \mathbf{Q},\ c \neq 0$$

so that

$$\max |ca_i|_v = |c|_v \max |a_i|_v ,$$

and therefore

$$H_{\mathbf{Q}}(ca_0, \ldots, ca_N) = H_{\mathbf{Q}}(a_0, \ldots, a_N) .$$

Multiplying by a common denominator, and factoring out a greatest common divisor, we can always select coordinates for the point which are relatively prime integers. If $(a_0, \ldots, a_N)$ are such coordinates, then we see that

$$H_{\mathbf{Q}}(P) = \max |a_i| ,$$

in other words, the height is the maximum of the ordinary absolute value of the coordinates.

If $a \in \mathbf{Q}$ we define

$$H_{\mathbf{Q}}(a) = H_{\mathbf{Q}}(1, a) ,$$

where $(1, a)$ are viewed as coordinates of a point in $\mathbf{P}^1$. Then it is immediate that if $a \neq 0$,

$$H_{\mathbf{Q}}(a) = H_{\mathbf{Q}}(1/a) .$$

Furthermore, if $a = b/d$ is an expression for a in lowest form as a quotient of integers, and $d > 0$, then

$$H_{\mathbf{Q}}(a) = \max (1, |a|) \cdot d .$$

Thus the height is expressed as a product of a factor at infinity, and a denominator for a.

Next let K be a number field, i.e. a finite extension of $\mathbf{Q}$. We let M_K be the set of absolute values on K extending those in $M_{\mathbf{Q}}$. From elementary algebraic number theory, we know that these are again of two types. Those at infinity, i.e. corresponding to the absolute value obtained by embedding K in the complex numbers, and those arising from prime ideals $\mathfrak{p}$ in the ring of algebraic integers $\mathfrak{o}_K$. For the moment, the reader need only know the most elementary facts as described for instance in my book *Algebra*, concerning absolute values in finite extensions. In particular, we know from this reference Chapter XII, § 3, that there exist positive integers n_v $(v \in M_K)$ such that for all $\alpha \in K$

$$\prod_{v \in M_K} |\alpha|_v^{n_v} = \prod_{v \in M_{\mathbf{Q}}} |N_{\mathbf{Q}}^K(\alpha)|_v = 1 .$$

Furthermore, if $v_0 \in M_{\mathbf{Q}}$ and $v|v_0$ denotes $v \in M_K$ extending v_0, then

$$\sum_{v|v_0} n_v = [K : \mathbf{Q}] .$$

We shall call n_v the **local degree** of v.

Let $P = (\alpha_0, \ldots, \alpha_N)$ be a point in $\mathbf{P}^N(K)$. We define

$$\boxed{H_K(P) = \prod_{v \in M_K} \max |\alpha_i|_v^{n_v}}$$

and we define the **absolute height**

$$\boxed{H(P) = H_K(P)^{1/[K:\mathbf{Q}]} .}$$

It is clear from the relation $\sum n_v = [K:\mathbf{Q}]$ that the absolute height is independent of the field K in which the coordinates α_i lie.

If the reader knows a minimum of algebraic number theory, e.g. as in Chapter I and II of my book on the subject, then he will immediately see that the other formulas for the height which we have given over the rational numbers also hold in a number field. Indeed, suppose the coordinates α_i are relatively prime algebraic integers. Let S_∞ be the set of absolute values at infinity in M_K. Then

$$H_K(P) = \prod_{v \in S_\infty} \max |\alpha_i|_v^{n_v} .$$

In other words, the height is described entirely in terms of the absolute values at infinity.

On the other hand, let $\alpha \in K$, and as before, let

$$\boxed{H_K(\alpha) = H_K(1, \alpha) = H_K(1/\alpha)}$$

(this last equality holding if $\alpha \neq 0$). Write an ideal factorization,

$$(\alpha) = \mathfrak{b}/\mathfrak{d} ,$$

where $\mathfrak{b}$, $\mathfrak{d}$ are relatively prime ideals in $\mathfrak{o}_K$. Then

$$\boxed{H_K(\alpha) = \prod_{v \in S_\infty} \max(1, |\alpha|_v^{n_v}) \cdot \mathbf{N}\mathfrak{d} ,}$$

where $\mathbf{N}$ denotes the absolute norm (cf. *Algebraic Number Theory*, Chapter II). Indeed, we have $\max(1, |\alpha|_v) > 1$ for a $\mathfrak{p}$-adic absolute value v if and only if $\mathfrak{p}$ divides the denominator $\mathfrak{d}$. In this case, let π be a prime element at $\mathfrak{p}$, and let $e = e(\mathfrak{p})$ and $f = f(\mathfrak{p})$ be the ramification index and residue class degree as usual. Then

$$|\pi^e|_\mathfrak{p} = |p|_p \quad \text{so} \quad |1/\pi|_\mathfrak{p} = p^{1/e} .$$

We have $n_\mathfrak{p} = e_\mathfrak{p} f_\mathfrak{p}$ so that

$$|1/\pi|_\mathfrak{p}^{n_\mathfrak{p}} = p^{f_\mathfrak{p}} = \mathbf{N}\mathfrak{p} .$$

This proves that $\mathbf{N}\mathfrak{d}$ is the contribution to the height from the non-archimedean absolute values.

A positive integer d is called a **denominator** for α if $d\alpha$ is an algebraic integer. We observe that $\mathbf{N}\mathfrak{d}$ is a denominator for α. Thus the height of α has been expressed as a product involving bounds for its absolute values at infinity and a "finite" component involving a denominator.

From the definition of the height as a product, it is obvious that if $\alpha_1, \ldots, \alpha_n$ are algebraic numbers in a field K, then

$$\boxed{H_K(\alpha_1 \cdots \alpha_n) \leqslant H_K(\alpha_1) \cdots H_K(\alpha_n) .}$$

For a single algebraic number α, we have an inequality in the opposite direction,

$$\boxed{H_K(\alpha^n) = H_K(\alpha)^n .}$$

In the preceding discussion we defined the height of points in projective space. Given a map of some set into projective space, we can then define the height relative to this mapping. Thus let

$$\varphi \colon A \to \mathbf{P}^N(\mathbf{Q}^a)$$

be a map of some set. We define for $P \in A$,

$$H_\varphi(P) = H(\varphi(P)) .$$

If A is an elliptic curve defined over some number field K_0, if φ is a rational map defined also over K_0, and $K \supset K_0$, we can define also the relative height

$$H_{K,\varphi}(P) = H_K(\varphi(P)) ,$$

for $P \in A_K$. The most common map φ is given by the x-coordinate,

$$x\colon A \to \mathbf{P}^1$$

into projective 1-space. In that case, we shall frequently write simply

$$H(P) \text{ instead of } H(x(P)) \text{ and } H_K(P) \text{ instead of } H_{K,x}(P) .$$

The same notation will apply when we define the logarithmic height below, omitting the x from the notation.

Lemma. *There is only a finite number of points of bounded height and bounded degree in* $\mathbf{P}^N(\mathbf{Q}^a)$.

Proof. It is easily seen that bounding the height yields a bound for the coefficients of the irreducible equation over $\mathbf{Z}$ for the coordinates of the points, because the coefficients of these equations are symmetric functions in the conjugates of these coordinates.

It will also be convenient to define the logarithmic height,

$$h_K(P) = \log H_K(P) , \quad \text{and} \quad h(P) = \log H(P) .$$

We note that the absolute heights H and h are defined on points in the algebraic closure $\mathbf{Q}^a$ of $\mathbf{Q}$, i.e. on $P^N(\mathbf{Q}^a)$.

Let $f_0, \ldots, f_{N'}$ be homogeneous polynomials of degree d, with coefficients in a number field, and in $N + 1$ variables $X_0, \ldots, X_N$. Assume that they have no common zero except the origin. Then these polynomials define a map

$$f\colon \mathbf{P}^N \to \mathbf{P}^{N'}$$

which associates to each point $P = (\alpha_0, \ldots, \alpha_N)$ the point

$$(f_0(P), \ldots, f_{N'}(P)) .$$

This is well defined because at least one coordinate $f_j(P) \neq 0$. Such a map will be called a **morphism of degree** d. If the coefficients of the polynomials f_j lie in K, then we say that f is **defined over** K.

The next theorem shows that up to a bounded function, the composite $h \circ f$ (height composed with the morphism) is dh.

Theorem 1.1. *Let $f: \mathbf{P}^N \to \mathbf{P}^{N'}$ be a morphism of degree d, defined over a number field. Then*

$$|h \circ f - dh|$$

is bounded on $\mathbf{P}^N(\mathbf{Q}^a)$.

Proof. Suppose f is defined over K_0. Trivial estimates using the triangle inequality show that for any point $P \in \mathbf{P}^N(K)$, where K contains K_0, we have

$$H_K(f(P)) \leqslant C_1^{[K:\mathbf{Q}]} H_K(P)^d .$$

Here C_1 is a constant depending only on f. Taking the $[K:\mathbf{Q}]$-th root and the log yields one inequality for $h \circ f - dh$. This is the "trivial" inequality.

The main point of the theorem is to prove an inequality on the other side. Since the polynomials $f_0, \ldots, f_N$ have no common zero except the origin, by the Hilbert Nullstellensatz, there exist polynomials $g_{ij} \in K_0[X_0, \ldots, X_N]$ and a positive integer m such that

$$X_i^{m+d} = \sum_j g_{ij} f_j .$$

Disregarding the monomials in g_{ij} of degree $\neq m$, we can assume without loss of generality that g_{ij} is homogeneous of degree m. It is also convenient to clear denominators, so we pick a positive integer $c > 0$ such that c is a denominator for all the coefficients of the polynomials g_{ij}. Multiplying by c, we may assume without loss of generality that we have the equation

$$cX_i^{m+d} = \sum_j g_{ij} f_j ,$$

where the coefficients of g_{ij} are algebraic integers in K_0. Let

$$x \in \mathbf{P}^N(K), \quad \text{and} \quad x = (x_0, \ldots, x_N) \quad \text{with } x_i \in \mathfrak{o}_K.$$

If $v \in M_K$ is not at infinity, and so is non-archimedean, then

$$|c|_v \, |x_i|_v^{m+d} \leqslant \max_j |f_j(x)|_v \, |x_i|_v^m$$

whence

$$|c|_v^{n_v} \max_i |x_i|_v^{n_v(m+d)} \leqslant \max_j |f_j(x)|_v^{n_v} \max_i |x_i|_v^{n_v m} .$$

If $v \in M_K$ is at infinity, then

$$|c|_v\, |x_i|_v^{m+d} \leqq (m+1)^N \max_{i,j}\; C_{ij} |x_i|_v^m \max_j |f_j(x)|_v ,$$

where C_{ij} are constants giving bounds for the absolute values at infinity for the coefficients of the polynomials involved. We then obtain

$$|c|_v^{n_v} \max_i |x_i|_v^{n_v(m+d)} \leqq C_2^{n_v} \max_i |x_i|_v^{mn_v} \max_j |f_j(x)|_v^{n_v} .$$

Taking the product yields

$$H_K(x)^{m+d} \leqq C_2^{[K:\mathbf{Q}]} H_K(x)^m H_K(f(x)) .$$

This proves the theorem.

Observe that in multiplicative notation, which is the way the theorem has been proved, it expresses the inequalities in the form

$$\boxed{C_2^{-[K:\mathbf{Q}]} H_K(x)^d \leqq H_K(f(x)) \leqq C_1^{[K:\mathbf{Q}]} H_K(x)^d .}$$

The constants C_1, C_2 depend only on f.

Example. We may view the elliptic curve A defined by the usual equation to be a curve in projective space $\mathbf{P}^2$, given in homogeneous coordinates by

$$Y^2 W = X^3 + aXW^2 + bW^3 .$$

Thus we have the homogeneous coordinates

$$W = X_0 , \qquad X = X_1 , \qquad Y = X_2 .$$

The function usually denoted by x on the curve may be viewed as defining a morphism into $\mathbf{P}^1$, so we write

$$x\colon A \to \mathbf{P}^1 ,$$

where $x(P)$ has coordinates $(1, x(P))$. If $2(x, y) = (x \circ 2, y \circ 2)$ then we recall that

$$x \circ 2 = \frac{x^4 - 2ax^2 - 8xb + a^2}{4(x^3 + ax + b)} .$$

Hence multiplication by 2 on the x-coordinate is induced by the morphism

$$f\colon \mathbf{P}^1 \to \mathbf{P}^1$$

given by the coordinates $f = (f_0, f_1)$ such that

$$f_0(w, x) = 4wx^3 + 4axw^3 + 4bw^4$$
$$f_1(w, x) = x^4 - 2ax^2w^2 - 8xbw^3 + a^2w^4 .$$

Similarly, $n(x, y) = (x \circ n, y \circ n)$, multiplication by n, is given by a morphism of degree n^2, obtainable in a similar way from the polynomials $\varphi_n, \psi_n^2, \psi_n^3, \omega_n$ arising from Theorem 2.1 of Chapter II. Applying Theorem 1.1 to this situation, we obtain:

Theorem 1.2. *Let A be an elliptic curve defined over a number field. For $P \in A(\mathbf{Q}^a)$ abbreviate*

$$h(P) = h(x(P)) .$$

Let n be a positive integer. Then

$$h(nP) = n^2 h(P) + O(1) ,$$

where $O(1)$ denotes a bounded function on $A(\mathbf{Q}^a)$.

Remark. The $O(1)$ in Theorem 1.2 seems to depend on n. However, we shall soon prove the stronger statement that $h(P)$ is in fact a quadratic form in P, plus a bounded function on $A(\mathbf{Q}^a)$. See Theorem 4.1. Hence the $O(1)$ depends only on the elliptic curve (the coefficients in its Weierstrass equation).

Before going into the quadraticity, we show how Theorem 1.2 applies to give what is known as the infinite descent. This will lead to the

Mordell–Weil Theorem. *If K is a number field, then $A(K)$ is finitely generated.*

§ 2. The Infinite Descent and Mordell–Weil Theorem

We axiomatize the situation. Let A be an abelian group. Assume given a function (which we shall call height),

$$h: A \to [0, \infty[,$$

having the following properties:

h 1. Given $P_0 \in A$, there exists a constant C_0 such that

$$h(P + P_0) \leqslant 2h(P) + C_0 .$$

h 2. There is an integer $m \geqslant 2$ and a constant C_1 such that

$$h(mP) \geqslant m^2 h(P) - C_1 .$$

h 3. There is only a finite number of points $P \in A$ with bounded height.

Theorem 2.1. *Under the above properties, if A/mA is finite, then A is finitely generated.*

Proof. Let $P_1, \ldots, P_r$ be representatives of A/mA. Given a point $Q_0 \in A$, we write

$$Q_0 = mQ_1 + P_{i_1}, \quad \text{so that } mQ_1 = P - P_{i_1}.$$

Inductively, we write

$$Q_n = mQ_{n+1} + P_{i_{n+1}}.$$

Then

$$-C_1 + m^2 h(Q_{n+1}) \leqslant h(mQ_{n+1}) \leqslant 2h(Q_n) + C_2$$

whence with $C_3 = C_1 + C_2$ we find

$$\begin{aligned} h(Q_{n+1}) &\leqslant \frac{2}{m^2} h(Q_n) + \frac{C_3}{m^2} \\ &\leqslant \frac{2^n}{m^{2n}} h(Q_1) + C_3(\delta + \delta^2 + \cdots) \end{aligned}$$

where $\delta = 2/m^2$. For large n, we see that Q_n lies in a set of bounded height. It is then clear that elements of this set, together with $P_1, \ldots, P_r$ generate A, as desired.

It is immediate from the addition formula that if A is an elliptic curve, and the abelian group consists of its algebraic points, then property **h 1** is satisfied. In the next section we shall see that h differs from a quadratic form by a bounded function, so property **h 1** also follows trivially from that. We already know Properties **h 2** and **h 3**. Consequently, in order to prove the Mordell–Weil theorem, all that remains to be done is to prove that A_K/mA_K is finite for some positive integer m. This will be done in a subsequent chapter since it involves different considerations from those of heights, namely Kummer theory in some form.

In the next section, we pursue the quadraticity property of the height.

§ 3. Quasi-Linear Algebra

Let A be an abelian group. Let L be a real valued function on $A \times A$. We call L **quasi-bilinear** if the function on $A \times A \times A$ given by

$$\Delta_1 L(P, P', Q) = L(P + P', Q) - L(P, Q) - L(P', Q)$$

is bounded, and similarly $\Delta_2 L$ is bounded.

A function h on A is called **quasi-quadratic** if h is even, i.e. $h(-P) = h(P)$, and the function Δh given by

$$\Delta h(P, Q) = h(P + Q) - h(P) - h(Q)$$

is quasi-bilinear.

We say that h **quasi-satisfies the parallelogram law** if h is even, and

$$h(P + Q) + h(P - Q) = 2h(P) + 2h(Q) + O(1) .$$

Lemma 1. *If h quasi-satisfies the parallelogram law, then h is quasi-quadratic, and if $L = \Delta h$, then*

$$h(P) - \tfrac{1}{2}L(P, P)$$

is bounded.

Proof. The usual argument works taking all relations mod $O(1)$. We have:

$$h(P + P' + Q) + h(P - P' + Q) = 2h(P + Q) + 2h(P') + O(1)$$

$$h(P + P' - Q) + h(P - P' - Q) = 2h(P - Q) + 2h(P') + O(1)$$

$$h(P + P' + Q) + h(P - P' - Q) = 2h(P' + Q) + 2h(P) + O(1)$$

$$h(P + P' - Q) + h(P - P' + Q) = 2h(P' - Q) + 2h(P) + O(1) .$$

Define

$$\langle P, Q \rangle = \tfrac{1}{4}[h(P + Q) - h(P - Q)] .$$

Taking the alternating sum of the four relations shows that this function of P, Q is quasi-bilinear. Since $h(2P) = 4h(P) + O(1)$, by the parallelogram law, we conclude that

$$h(P) = \langle P, P \rangle + O(1) .$$

It is then immediate that Δh is quasi-bilinear, and the lemma is then obvious.

Lemma 2. *If L is quasi-bilinear, then there exists a unique bilinear function L_0 such that $L - L_0$ is bounded. In fact,*

$$L_0(P, Q) = \lim_{n \to \infty} \frac{L(2^n P, 2^n Q)}{4^n} .$$

If h is quasi-quadratic, then there exists a unique quadratic function h_0 such that $h_0 - h$ is bounded. In fact,

$$h_0(P) = \lim_{n\to\infty} \frac{h(2^n P)}{4^n}.$$

Proof. Write

$$L_n(P, Q) = \frac{L(2^n P, 2^n Q)}{4^n}.$$

Suppose that C is a positive number such that

$$|\Delta_1(P, P', Q)| \quad \text{and} \quad |\Delta_2(P, Q, Q')| \leqslant C$$

for all P, P', Q, Q'. From quasi-bilinearity we find

$$|L_{n+1}(P, Q) - L_n(P, Q)| \leqslant 3C/4^n.$$

Hence $\{L_n(P, Q)\}$ is a Cauchy sequence, and it is immediately verified that the limit is bilinear.

We note that we get an explicit bound on $L - L_0$, namely $3C$, if we know a bound for the $O(1)$ expressing the quasi-bilinearity of L. The same holds for the quadratic case, with a similar low integer multiplying $O(1)$.

This averaging process was applied by both Néron and Tate, approximately simultaneously. Néron also applies it to local components of the height function [Ne 2].

More generally, one can perform a similar averaging procedure on any map, due to Tate. We give the result even though we shall not need it.

Lemma 3. *Let S be a set and $f: S \to S$ a map of S into itself. Let $h: S \to \mathbf{R}$ be a function. Assume that there exists a positive integer d such that*

$$h \circ f - dh \text{ is bounded}.$$

Then there exists a unique function h_f such that:

(i) *$h_f - h$ is bounded.*
(ii) *$h_f \circ f = dh_f$.*

Proof. If h'_f and h''_f are two such functions, then $h'_f - h''_f$ is bounded, and it is clear from (ii) that $h'_f - h''_f = 0$, so uniqueness is obvious. As to existence, we let

$$h_f(x) = \lim_{n\to\infty} \frac{1}{d^n} h(f^n(x)).$$

If $|h \circ f - dh| \leqslant C$, then

$$\left|\frac{1}{d^{n+1}} h \circ f^{n+1}(x) - \frac{1}{d^n} h \circ f^n(x)\right| \leqslant C/d^{n+1}.$$

Hence

$$\left|\frac{1}{d^{n+k}} h \circ f^{n+k}(x) - \frac{1}{d^n} h \circ f^n(x)\right| \leqslant \frac{C}{d^n} \frac{1}{1 - 1/d}.$$

Hence the sequence $\left\{\frac{1}{d^n} h \circ f^n\right\}$ is Cauchy, its limit exists and trivially satisfies (i) and (ii), as was to be shown.

The lemmas give the existence of a unique function in a class of functions differing by bounded functions. It will therefore be useful to use the terminology that two functions are **equivalent** if their difference is bounded. Lemmas 1 and 2 show that a function which quasi-satisfies the parallelogram law is equivalent to a quadratic form.

§ 4. Quadraticity of the Height

We return to the elliptic curve A defined over a number field by

$$y^2 = x^3 + ax + b.$$

We let h be the absolute (logarithmic) height, defined by

$$h(P) = h(x(P)), \quad P \in A(\mathbf{Q}^a).$$

Theorem 4.1. *The function h quasi-satisfies the parallelogram law, i.e.*

$$h(P + Q) + h(P - Q) = 2h(P) + 2h(Q) + O(1).$$

Consequently h is equivalent to a unique quadratic function.

Proof. Let x_1, x_2, x_3, x_4 be the x-coordinates of the points $P, Q, P + Q, P - Q$. Then from the addition formula, one gets as a classical exercise:

$$x_3 + x_4 = \frac{2(x_1 + x_2)(a + x_1 x_2) + 4b}{(x_1 - x_2)^2}$$

$$x_3 x_4 = \frac{(x_1 x_2 - a)^2 - 4b(x_1 + x_2)}{(x_1 - x_2)^2}.$$

There is a natural mapping

$$\sigma : \mathbf{P}^1 \times \mathbf{P}^1 \to \mathbf{P}^2$$

given by

$$(w_1, x_1) \times (w_2, x_2) \overset{\sigma}{\mapsto} (w_1 w_2, w_1 x_2 + w_2 x_1, x_1 x_2) .$$

On affine coordinates, this mapping yields

$$(1, x_1) \times (1, x_2) \mapsto (1, x_1 + x_2, x_1 x_2) .$$

Let $x^{(2)}(P, Q) = (x(P), x(Q))$. We then obtain a composite map

$$\sigma \circ x^{(2)} : A \times A \to \mathbf{P}^2$$

by

$$(P, Q) \mapsto \sigma(x(P), x(Q)) .$$

Let

$$g : \mathbf{P}^2 \to \mathbf{P}^2$$

be the map given on homogeneous coordinates (t, u, v) by

$$g(t, u, v) = (u^2 - 4tv, 2u(at + v) + 4bt^2, (at - v)^2 - 4btu) .$$

Then g is a morphism of degree 2. The formulas for $x_3 + x_4, x_3 x_4$ show that there is a commutative diagram

$$\begin{array}{ccc} A \times A & \longrightarrow & A \times A \\ {\scriptstyle \sigma \circ x^{(2)}} \downarrow & & \downarrow {\scriptstyle \sigma \circ x^{(2)}} \\ \mathbf{P}^2 & \underset{g}{\longrightarrow} & \mathbf{P}^2 \end{array}$$

given on elements by

$$\begin{array}{ccc} (P, Q) & \longmapsto & (P + Q, P - Q) \\ \downarrow & & \downarrow \\ \sigma(x(P), x(Q)) & \underset{g}{\mapsto} & \sigma(x(P + Q), x(P - Q)) \end{array}$$

or in other words,

$$g(\sigma(x(P), x(Q))) = \sigma(x(P+Q), x(P-Q)).$$

We know from Theorem 1.1 that $h \circ g - 2h = O(1)$. Hence

$$h(\sigma(x(P+Q), x(P-Q)) = 2h(\sigma(x(P), x(Q)) + O(1).$$

The proof of Theorem 4.1 will therefore be concluded by the following lemma.

Lemma. $$h(\sigma(x(P), x(Q))) = h(P) + h(Q) + O(1).$$

Proof. This amounts to showing that for α, β in a number field K we have

$$h(1, \alpha+\beta, \alpha\beta) = h(\alpha) + h(\beta) + O(1),$$

or in multiplicative notation,

$$H(1, \alpha+\beta, \alpha\beta) \gg\ll H(\alpha)H(\beta).$$

The height is a product taken over non-archimedean absolute values, and absolute values at infinity. For the non-archimedean v, we actually get an equality,

$$\max(1, |\alpha+\beta|_v, |\alpha\beta|_v) = \max(1, |\alpha|_v) \max(1, |\beta|_v)$$

as is easily seen by distinguishing cases when $|\alpha|_v \leqslant 1$ or $\geqslant 1$, and similarly for $|\beta|_v$. For v at infinity, one gets quasi-multiplicativity, with the usual constants. The estimates are trivial and left to the reader.

In the light of the quasi-linear algebra of the preceding section, there is a unique bilinear form $\langle\ ,\ \rangle$ on the group of algebraic points on A such that

$$h(P) = \tfrac{1}{2}\langle P, P\rangle + O(1).$$

This quadratic or bilinear form is called the **Néron–Tate form**. We denote

$$\hat{h}(P) = \tfrac{1}{2}\langle P, P\rangle,$$

and call $\hat{h}$ the **Néron–Tate height**.

Theorem 4.2. *Let* $|\ |$ *be the norm associated with the Néron–Tate form*, i.e.

$$|P| = \sqrt{\langle P, P\rangle}.$$

Then $|P| = 0$ *if and only if* P *is a torsion point.*

Proof. For any positive integer n we have

$$|nP| = n|P| ,$$

whence the assertion is obvious, using the fact that there is only a finite number of points of bounded height in a given number field.

Let K be a number field. The group A_K is finitely generated. If we take the tensor product $\mathbf{R} \otimes A_K$, then we obtain a finite dimensional vector space over the reals, whose dimension is equal to the rank of A_K (the torsion group disappears). The quadratic form extends naturally to this real space, and A_K may be viewed as a lattice in $\mathbf{R} \otimes A_K$. In the light of Theorem 4.2 it was believed to be automatic that the quadratic form on $\mathbf{R} \otimes A_K$ is positive definite. Cassels pointed out that the matter is not quite so automatic, one needs to use Minkowski's elementary theorem that a convex set which is symmetric about the origin contains a non-zero lattice point if its volume is sufficiently large. (The reader will find a proof in almost any book on algebraic number theory.) The following is then true.

Theorem 4.3. *Let V be a finite dimensional vector space over* $\mathbf{R}$, *and let Λ be a lattice in V. Let h be a quadratic form on V. Assume:*

(i) *If P is in the lattice and $P \neq O$ then $h(P) > 0$.*
(ii) *There is only a finite number of $P \in \Lambda$ with bounded $h(P)$.*

Then h is positive definite on V.

Proof. There exists a basis of V such that for any point X in V with coordinates $(x_1, \ldots, x_n)$ with respect to this basis we have

$$h(X) = \sum_{i=1}^{r} x_i^2 - \sum_{i=r+1}^{m} x_i^2 , \qquad m \leqslant \dim V.$$

If we write $X = (X', X'', X^*)$, where X' is the projection of X on the subspace where h is positive definite, X'' the projection where h is negative definite, and X^* the projection where h is a null form, then

$$h(X) = X'^2 - X''^2 .$$

Let b, $B > 0$. The inequalities

$$X'^2 < b \quad \text{and} \quad X''^2 < B$$

define a symmetric convex set in V. Taking b small and B large, we conclude from (i) and Minkowski's theorem that h is positive semidefinite on V. Again taking b small and allowing large coordinates for X^*, we conclude from (ii) and Minkowski's theorem that h has to be positive definite on V, as desired.

The investigation of the lattice A_K (modulo torsion) is one of the fundamental problems of diophantine analysis on elliptic curves.

To begin with, can one say anything about the **successive minima**? We recall that these minima are obtained as follows. We select a point P_1 of minimum height $\neq 0$ (the height is here the Néron–Tate height). We then let P_2 be a point linearly independent from P_1 of height minimum $\geqslant \hat{h}(P_1)$. We continue in this manner to obtain a maximum set of linearly independent elements $P_1, \ldots, P_r$ called the **successive minima**.

We suppose that A is defined by a Weierstrass equation

$$y^2 = x^3 + ax + b\,,$$

and we assume for simplicity that $K = \mathbf{Q}$, so that we take $a, b \in \mathbf{Z}$. We replace A by a curve isomorphic to A over $\mathbf{Q}$, also having integral coefficients, but such that its discriminant has minimal absolute value. Such a curve is called a minimal model. One can always find a minimal model because $\mathbf{Z}$ has unique factorization into primes.

It seems a reasonable guess that uniformly for all such minimal models of elliptic curves over $\mathbf{Z}$, one has

$$\hat{h}(P_1) \gg \log |\Delta|\,.$$

It is difficult to guess how the next successive minima may behave. Do the ratios of the successive heights remain bounded by a similar bound involving $|\Delta|$, or possibly a power of $\log |\Delta|$, or can they become much larger? In other words, does the picture of the successive minima look like one or the other of the following two figures?

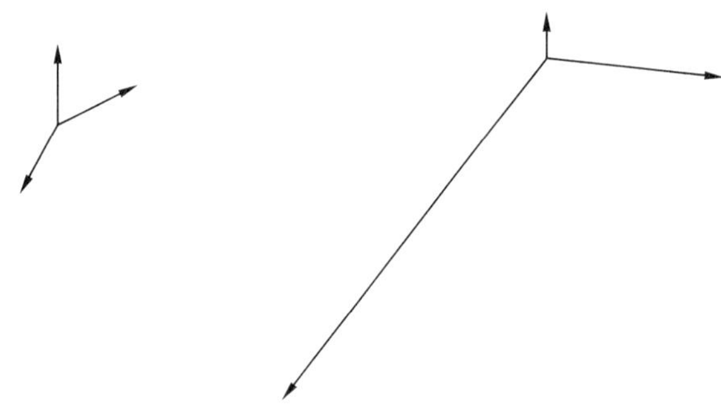

Fig. 6

Since one has no examples of elliptic curves with arbitrarily high rank, and since it is generally difficult to find generators for the Mordell–Weil group, the answer to these questions appears today to be exceedingly hard.

Over a number field K, it is not possible to construct a global minimal model, but one can work with quasi-minimal models, making isomorphisms to reduce the prime factorization of Δ as much as possible. Such models are not uniquely determined, but there is only a finite number of them in a given isomorphism class over K.

Manin [Man 4] gives a splendid discussion relating the regulator of a basis for the Mordell–Weil group and Birch–Swinnerton Dyer conjecture. The questions raised above are obviously connected with the value of the zeta function of the elliptic curve at 1, a topic which we have not mentioned at all in this book as it takes substantial space to develop.

§ 5. Linear Dependence of Algebraic Points

Both on the multiplicative group and on an elliptic curve it is important to know effective bounds for the coefficients of a linear relation among a finite number of elements, in terms of the heights of these elements. We begin with the multiplicative group. The method used in the next theorem already occurs in Stark [St 3], who applies it to units. For a refinement using the geometry of numbers, cf. [VdP–L 2].

Theorem 5.1. *Let K be a number field. Let $\alpha_1, \ldots, \alpha_r$ be multiplicatively independent non-zero elements of K. Let N be the period of an element $\alpha \neq 0$ in K with respect to the group generated by $\alpha_1, \ldots, \alpha_r$ and K^*_{tor} (=roots of unity in K). Then N satisfies the bound*

$$\phi(N) \ll (h_1 + \cdots + h_r)^r$$

where $h_j = \log H_K(\alpha_j)$, and the constant implicit in $\ll$ depends only on $[K : \mathbf{Q}]$.

Proof. We write

$$\alpha_1^{m_1} \cdots \alpha_r^{m_r} \zeta = \alpha^N ,$$

where ζ is a root of unity. We may assume that

$$(m_1, \ldots, m_r, N) = 1 .$$

By Dirichlet's box principle applied to

$$(qm_1/N, \ldots, qm_r/N) \pmod{\mathbf{Z}^r} ,$$

we can find an integer q prime to N with $0 < q < N$ such that

$$|qm_j/N - s_j| \ll \phi(N)^{-1/r}$$

with appropriate integers s_j. Let

$$n_j = qm_j - Ns_j$$

so that

$$|n_j| \ll N\phi(N)^{-1/r}.$$

We obtain

$$\alpha_1^{n_1} \dots \alpha_r^{n_r} \zeta_0 = \beta^N$$

for some $\beta \in K^*$ (not a root of unity) and some root of unity ζ_0. Hence

$$H_K(\beta)^N = H_K(\beta^N) \leqslant (H_1 \cdots H_r)^n$$

where $n = \max |n_j|$. Since the heights of elements of K which are not roots of unity are bounded from below by a positive number depending only on the degree $[K : \mathbf{Q}]$, this concludes the proof.

Observe how the trivial multiplicative property

$$H_K(\beta^N) = H_K(\beta)^N$$

for a single element β is used, and the submultiplicativity for a product of several elements. In the analogous situation on elliptic curves, we shall use the Néron–Tate quadratic form. It is relevant here to recall the Lehmer conjecture:

There is a constant $C_0 > 1$ such that, if α is an algebraic number in a number field K, with $\alpha \neq$ root of unity, then

$$H_K(\alpha) \geqslant C_0 .$$

See Lehmer, Ann. of Math. **34** (1933) pp. 461–479. For a recent reference to this conjecture, see Blanskby–Montgomery [B–M].

Theorem 5.2. *Let $\alpha_1, \dots, \alpha_{r+1}$ be non-zero elements of K. Assume that they have multiplicative rank r. Then there exists a multiplicative relation*

$$\alpha_1^{m_1} \dots \alpha_{r+1}^{m_{r+1}} = 1, \qquad (\textit{not all } m_j = 0),$$

with $M = \max |m_j|$ satisfying

$$\phi(M) \ll (h_1 + \cdots + h_{r+1})^r.$$

Proof. We omit it, but we shall give the full details of the analogous isomorphic proof in the case of elliptic curves.

We now give an application of Theorem 5.2. Let $\alpha_1, \dots, \alpha_r$ again be multiplicatively independent. Let α_0 be a generator for the group of roots of unity in K. Let Γ

be the multiplicative group generated by $\alpha_1 \ldots, \alpha_r$, and let Γ' be the division group of Γ in K^*, i.e. the group of all elements α in K^* such that $\alpha^m \in \Gamma$ for some positive integer m. The torsion subgroup of Γ' is again generated by α_0, and the rank of Γ' is r. The group Γ' is important in connection with Kummer theory, where we want to analyse Kummer extensions

$$K(\Gamma^{1/n}) \quad \text{or} \quad K(\Gamma'^{1/n}),$$

both when the n-th roots of unity are contained in K, and when they are not. See Chapter V, § 4.

We want to choose the generators α'_j to have heights bounded in terms of $H_K(\alpha_j)$. Note that $H_K(\alpha_0) = 1$. We assume that

$$H_K(\alpha_1) \leqslant H_K(\alpha_2) \leqslant \cdots \leqslant H_K(\alpha_r).$$

Corollary. *There exist free generators α'_j $(j = 1, \ldots, r)$ as above satisfying*

$$\log H_K(\alpha'_j) \leqslant \sum_{\nu=1}^{j} \log H_K(\alpha_\nu) \quad \textit{for } j = 1, \ldots, r.$$

Let $\alpha'_0 = \alpha_0$. For each k there exists a relation

$$m \log \alpha_k = \sum_{j=0}^{r} m_j \log \alpha'_j$$

with integers m, m_j and $m > 0$, such that if $M = \max(m, |m_j|)$, then

$$\phi(M) \leqslant C_2(\log H_K(\alpha_r))^r.$$

Here C_1, C_2, are effectively computable constants depending only on r and $[K : \mathbf{Q}]$. The logs are principal valued.

Proof. This is obvious by the usual method of obtaining a basis for a sublattice from the basis of a lattice in elementary linear algebra, combined with the effective bounds on periods of elements given by Theorem 5.2. We reproduce the arguments briefly for the convenience of the reader. Let n be the index of Γ in Γ'. Then Γ'^n is a subgroup of finite index in Γ. Let $n_{j,j}$ be the smallest positive integer such that there exist integers $n_{j,0}, \ldots, n_{j,j-1}$ satisfying

$$\alpha_0^{n_{j,0}} \ldots \alpha_j^{n_{j,j}} \in \Gamma'^n.$$

Without loss of generality, we may assume $0 \leqslant n_{j,k} \leqslant n - 1$. Let

$$\alpha_0^{n_{j,0}} \ldots \alpha_j^{n_{j,j}} = \alpha_j'^n.$$

Then α'_j ($j = 1, \ldots, r$) form a basis of Γ' modulo roots of unity. Furthermore,

$$H_K(\alpha'_j)^n = H_K(\alpha'^n_j) \leqslant H_K(\alpha_1)^{n_{j,1}} \cdots H_K(\alpha_j)^{n_{j,j}} .$$

Taking the n-th root yields a bound $H_K(\alpha'_j) \leqslant H_K(\alpha_1) \cdots H_K(\alpha_j)$. By Theorem 5.2, for each k we also have a relation

$$\alpha_k^m = \alpha'^{m_1}_1 \cdots \alpha'^{m_r}_r$$

with integers $m, m_1, \ldots, m_r$ and $m > 0$ satisfying

$$\phi(M) \ll (\max \log H_K(\alpha^j), \log H_K(\alpha'_j))^r ,$$

which we combine with the previous estimate to conclude the proof.

Estimates as in the corollary are used in the diophantine approximation proofs, see Cijsouw–Waldschmidt [C–W], and Chapter XI.

We now pass to the analogous theory on elliptic curves.

Let K be a number field, and let A be an elliptic curve defined over K. Let $P_1, \ldots, P_n \in A_K$. These points may be linearly dependent (over $\mathbf{Z}$). We ask for a bound on the coefficients of a linear relation. We know from Theorem 1.4 of Chapter III that the group of torsion points $A_{K,\mathrm{tor}}$ is finite, because this is even true for the group of points of A in a $\mathfrak{p}$-adic completion of K, and the proof gave an effective bound for its order.

In a theorem of Ribet–Bashmakov [Ri], [Ba], it was important to know the answer to the following question: If $P_1, \ldots, P_r$ are linearly independent, then they remain linearly independent in $A_K/lA_K \bmod \mathbf{Z}/l\mathbf{Z}$ for all sufficiently large primes l, and an effective bound for the l when they are dependent can be given. The Mordell–Weil theorem as we have proved it does not give such a bound. Cassels gave a simple proof using the Dirichlet box principle and a height inequality along Stark's lines. The same idea of proof can be used to give a bound for the period of a point with respect to $\{P_1, \ldots, P_r\}$, and we proceed as follows.

Theorem 5.3. *Let $P_1, \ldots, P_r$ be points in A_K, linearly independent over $\mathbf{Z}$. Let Γ be the group generated by these points and the torsion subgroup of A_K. Let $Q \in A_K$ be such that some positive multiple of Q lies in Γ. Then the exact period N of Q with respect to Γ satisfies the bound*

$$\phi(N) \leqslant C(|P_1| + \cdots + |P_r|)^r$$

where $|P|$ is the norm arising from the Néron–Tate quadratic form, and C is a constant depending only on A and the degree of K over $\mathbf{Q}$.

Proof. Suppose we have a linear relation

$$m_1 P_1 + \cdots + m_r P_r \equiv NQ \pmod{A_{K,\mathrm{tor}}}$$

and not all m_j are 0. Let $d = (m_1, \ldots, m_r, N)$. Then write

$$m_j = dm'_j, \qquad N = dN'$$

with $(m'_1, \ldots, m'_r, N') = 1$. Then for some torsion point P_0 we get

$$m'_1 P_1 + \cdots + m'_r P_r + P_0 = N'Q.$$

Hence we may assume without loss of generality that to start with,

$$(m_1, \ldots, m_r, N) = 1,$$

and we have a relation

$$m_1 P_1 + \cdots + m_r P_r + P_0 = NQ,$$

where P_0 is a torsion point.

By Dirichlet's box principle applied to multiples

$$(q\alpha_1, \ldots, q\alpha_r) \ (\mathrm{mod}\ \mathbf{Z}^r)$$

with $\alpha_j = m_j/N$, we conclude that there exists an integer q prime to N with $0 < q < N$ such that

$$|qm_j/N - s_j| \ll \phi(N)^{-1/r}$$

with appropriate integers s_j. Let

$$n_j = qm_j - Ns_j,$$

so that

$$|n_j| \ll N\phi(N)^{-1/r}.$$

We obtain

$$n_1 P_1 + \cdots + n_r P_r + P_0 = NQ_1$$

for some $Q_1 \in A_K$. Hence we find

$$CN \leqslant N\phi(N)^{-1/r}(|P_1| + \cdots + |P_r|),$$

where C is a number >0 depending only on the minimal height of points of A_K. (Recall that there is only a finite number of points in projective space over K of bounded height.) This proves the theorem.

Remark. In any book on elementary number theory, or book about prime numbers, the reader will find that the Euler function satisfies a bound

$$\phi(N) \gg N/\log\log N .$$

This gives an explicit bound for N itself in terms of the heights of the points P_j. If N is a prime power, then $\phi(N) \gg\ll N$.

Suppose that P_{r+1} is not a torsion point but is linearly dependent on $P_1, \ldots, P_r$. Then any r elements among $P_1, \ldots, P_{r+1}$ are linearly independent mod $A_{K,\mathrm{tor}}$, and the points $P_1, \ldots, P_{r+1}$ have rank r (mod $A_{K,\mathrm{tor}}$). Thus there is a unique relation

$$m_1 P_1 + \cdots + m_{r+1} P_{r+1} \equiv O \ (\mathrm{mod}\ A_{K,\mathrm{tor}})$$

with integers m_i not all 0 such that $(m_1, \ldots, m_{r+1}) = 1$. We can apply Theorem 5.3 to each coefficient m_i. We then find:

Theorem 5.4. *Let $P_1, \ldots, P_{r+1} \in A_K$ have rank r. Then there exists a relation of linear dependence*

$$m_1 P_1 + \cdots + m_{r+1} P_{r+1} = O$$

with integers m_i not all 0, *such that if $M = \max |m_i|$, then*

$$\phi(M) \ll (|P_1| + \cdots + |P_{r+1}|)^r .$$

The constant in $\ll$ depends only on A and the degree $[K : \mathbf{Q}]$.

Remark. In all the above, we have uniformity for all fields K of bounded degree because of two facts:

(i) The number of points in projective space of bounded height and bounded degree is finite.

(ii) The order of the torsion group $A_{K,\mathrm{tor}}$ is uniformly bounded for all K of bounded degree.

For this latter fact, we select a prime p. The p-adic field $\mathbf{Q}_p$ has only a finite number of extensions of bounded degree, and we know that in any such extension, the order of the torsion group is bounded. Hence for any prime ideal $\mathfrak{p}$ in K over p, the torsion group of A in $K_{\mathfrak{p}}$ is bounded.

§ 6. Local Decomposition of the Height

Suppose A is defined over the number field K. Let M_K as usual be the set of normalized absolute values on K, and for $P \in A_K$ define

$$\lambda_K(P) = \sum n_v \lambda_v(P) \quad \text{if } P \neq O$$
$$\lambda_K(O) = 0 .$$

The sum is taken for $v \in M_K$, and n_v is the local degree, so that the product formula is valid. Each λ_v is the local Néron function on A_{K_v}.

For all but a finite number of v, we know from Chapter III, that

$$\lambda_v(P) = \tfrac{1}{2} \log \max \{1, |x(P)|_v\} = \tfrac{1}{2} h_v(P).$$

For the other v, we also know that $\lambda_v - \frac{1}{2}h_v$ is bounded. Hence

$$\lambda_K - \tfrac{1}{2}h_K \text{ is bounded}.$$

We define the absolute function

$$\lambda = \frac{1}{[K:\mathbf{Q}]} \lambda_K,$$

and it then follows that $\lambda - \frac{1}{2}h$ is bounded. In fact, suppose that A is defined over a number field K_0 and that $K \supset K_0$. We have an inequality by Chapter I, Theorem 8.4 and Chapter III, Theorem 4.5, namely for each absolute value v on K,

$$|\lambda_v - \tfrac{1}{2}h_v| \leqslant C(v_0)$$

where v_0 is the restriction of v to K_0, and $C(v_0) = 1$ for all but a finite number of v_0, for instance if v_0 is non-archimedean, not dividing 2, 3, and such that A has good reduction at v_0. Hence

$$\begin{aligned} |\lambda_K - \tfrac{1}{2}h_K| &= |\sum n_v \lambda_v - \tfrac{1}{2} \sum n_v h_v| \\ &\leqslant \sum_{v_0} \sum_{v|v_0} n_{v/v_0} C(v_0), \end{aligned}$$

and it then follows that $\lambda - \frac{1}{2}h$ is bounded.

The function λ satisfies the quadratic relation

$$\lambda(P+Q) + \lambda(P-Q) = 2\lambda(P) + 2\lambda(Q)$$

for all pairs P, Q such that P, Q, $P \pm Q \neq O$. If P or $Q = O$ then the definition $\lambda(O) = 0$ also makes this relation true. *It is in fact true in all cases.* Suppose $P \neq O$ and $Q \neq O$ but $P + Q = O$ so $P = -Q$. The relation then amounts to

$$\lambda(2P) = 4\lambda(P).$$

Formally, this means that we have to show that the formal linear combination

$$(2P) - 4(P)$$

can be expressed in terms of "parallelograms" on which λ vanishes. In fact, denote formally

$$S(P, Q) = (P + Q) + (P - Q) - 2(P) - 2(Q)\,.$$

Then as in Kubert [K], we have

$$2(2P) - 8(P) = S(Q - 2P, P) + 2S(Q - P, P) + S(Q, P) - S(Q - P, 2P)\,.$$

We select Q of infinite order, independent of P. Then λ vanishes on each parallelogram of the right-hand side, whence λ vanishes on the linear combination on the left, as desired. Thus λ is unrestrictedly quadratic.

Let $\hat{h}$ be the quadratic function such that $h - \hat{h}$ is bounded. Since λ is a quadratic function and $\lambda - \frac{1}{2}h$ is bounded, we obtain:

Theorem 6.1. $\lambda = \frac{1}{2}\hat{h}$.

Thus the global Néron function has been expressed as a sum of local components.

To construct the quadratic form differing from $\frac{1}{2}h$ by a bounded function, we could have omitted entirely the global considerations of § 3 and § 4. It suffices to have carried out the local theory at each v, as in Chapter I, Theorem 8.4 and Chapter III, Theorem 4.5. On the other hand, it is shorter for many applications which do not need such refined results to deal directly with the global arguments and to omit this more elaborate local theory.

Chapter V. Kummer Theory

In this chapter, we investigate the nature of the extension obtained by extracting m-th roots of rational points. More precisely, let A be an elliptic curve defined over the number field K. Let $Q \in A_K$. We look into the field $K(P)$, where P is some point such that $mP = Q$, where m is a positive integer. Let A_m as usual denote the group of points of period m on A. We shall first assume that $A_m \subset A_K$. We give a computational proof that $A_K/2A_K$ is finite, making explicit use of the duplication formulas for points on A. This is the proof given by Weil [We 1], eight years after Mordell gave his first proof of the finite generation of A_K, see also the account given in Mordell's book [Mo]. Next we give a second proof depending on more algebraic number theory and reduction mod various primes, but involving no computations and also applicable to abelian varieties.

Finally, we give Bashmakov's result concerning the Galois group of division points, stating that it is almost always as large as the a priori structure of the situation will allow. This was generalized to abelian varieties with complex multiplication by Ribet [Ri]. The exposition here owes to both. However we do not limit ourselves to prime level, with an eye to possible future applications. The Bashmakov theorem will be used in Chapter IX. For another use, cf. Lang–Trotter [L–T].

§ 1. $A_K/2A_K$ is Finite

Let K be a number field and let A be an elliptic curve defined over K. Making a finite extension of K if necessary, we may assume that $A_2 \subset A_K$, in other words the 2-torsion points are rational over K. If A is in Weierstrass form

$$y^2 = (x - \alpha_1)(x - \alpha_2)(x - \alpha_3) = h(x) ,$$

this means that all three roots α_i lie in K. Let

$$Q_i = (\alpha_i, 0) .$$

Then Q_1, Q_2, Q_3 are the points of order 2 other than O.

Theorem 1.1. *The factor group $A_K/2A_K$ is finite.*

The proof will be carried out by constructing homomorphisms

$$\theta_i\colon A_K \to K^*/K^{*2}$$

such that

$$\bigcap_{i=1}^{3} \operatorname{Ker} \theta_i \subset 2A_K .$$

If we can then prove that the image of θ_i is contained in a finitely generated group, whence a finite group since every element of the image has period 2, it follows that $A_K/2A_K$ is finite. We do this by a sequence of lemmas.

We do not need the properties of number fields all the way. The first arguments depend only on field properties, and that K has characteristic $\neq 2, 3$ would suffice.

We define θ_i by letting:

$$\begin{aligned}
&\theta_i(O) = 1 ;\\
&\theta_i(x, y) = x - \alpha_i \bmod K^{*2} \quad \text{if } x \neq \alpha_i ;\\
&\theta_i(\alpha_i, 0) = (\alpha_i - \alpha_j)(\alpha_i - \alpha_k) \bmod K^{*2} ;
\end{aligned}$$

where α_j, α_k are the other two roots of h.

Lemma 1. *$\theta_i\colon A_K \to K^*/K^{*2}$ is a homomorphism.*

Proof. With the obvious notation, we can write θ_α for the above homomorphism sending $(x, y) \mapsto x - \alpha$ if $y \neq 0$. It is immediately verified that θ_α is a homomorphism on the group of points of order 2. We let the reader show that it satisfies the homomorphic property for the sum of two points, one of which is of order 2. Suppose we now have three points $P_i = (x_i, y_i)$ with $i = 1, 2, 3$ such that $y_i \neq 0$, and such that

$$P_1 + P_2 + P_3 = O .$$

This means that the three points lie on the same straight line

$$y_i = \lambda x_i + \mu .$$

Then $x_i - \alpha$ $(i = 1, 2, 3)$ are the roots of the polynomial

$$h(x + \alpha) - (\lambda(x + \alpha) + \mu)^2 ,$$

whose constant term is $-(\lambda\alpha + \mu)^2$. Hence

$$(x_1 - \alpha)(x_2 - \alpha)(x_3 - \alpha) = (\lambda\alpha + \mu)^2 ,$$

thereby proving the lemma.

Lemma 2. $\bigcap \operatorname{Ker} \theta_i \subset 2A_K$.

Proof. We have to show that if $(\xi, \eta) \in A_K$ and $\xi - \alpha_i$ is a square in K^* for $i = 1, 2, 3$ then

$$(\xi, \eta) = 2(x, y)$$

for some $(x, y) \in A_K$. This depends on some general remarks concerning multiplication by 2 on A which we now make.

Suppose we have a generic point (x, y) on A. Let

$$2(x, y) = (x_2, y_2).$$

Then general theoretical reasons which will be apparent later from Kummer theory imply that $x_2 - \alpha_i$ is the square of a rational function in (x, y). If A is defined by the usual equation

$$y^2 = x^3 + ax + b,$$

then a direct computation yields the formula, for $\alpha = \alpha_i$ $(i = 1, 2, 3)$:

Lemma 3.

$$\boxed{x_2 - \alpha = \left(\frac{x^2 - a - 2\alpha x - 2\alpha^2}{2y}\right)^2}$$

so that the computation avoids any general theory for the moment.

Let us suppose that $x_2 - \alpha_i$ is a square in K, say

$$x_2 - \alpha_i = z_i^2, \quad z_i \in K.$$

The system of equations

$$u + v\alpha_i + w\alpha_i^2 = z_i \quad \text{for } i = 1, 2, 3$$

has a unique solution with u, v, w in K, because the determinant of its coefficients is Vandermonde. We then obtain

$$(u + v\alpha_i + w\alpha_i^2)^2 = x_2 - \alpha_i.$$

We multiply out, using the relations

$$\alpha^3 + a\alpha + b = 0 \quad \text{and} \quad \alpha^4 + a\alpha^2 + b\alpha = 0,$$

valid for $\alpha = \alpha_i$. Using again the uniqueness of a system of linear equations with a Vandermonde system of coefficients, this means that the above relation amounts to the three relations

$$(*)\qquad \begin{aligned} u^2 - 2vwb &= x_2 \\ 2uv - 2vwa - w^2b &= -1 \\ v^2 + 2uw - w^2a &= 0\,. \end{aligned}$$

Eliminating u from the second and third equation, we obtain

$$v^3 + avw^2 + bw^3 = w\,.$$

If $w = 0$ then $v = 0$ and we cannot have $u^2 = x - \alpha_i$ for all i. Hence $w \neq 0$. Dividing by w^3 yields

$$\left(\frac{v}{w}\right)^3 + a\left(\frac{v}{w}\right) + b = \frac{1}{w^2}.$$

We now let

$$x_1 = v/w \quad \text{and} \quad y_1 = 1/w\,.$$

Then (x_1, y_1) is in A_K! From the third equation in (*) we get

$$2uy_1 + x_1^2 - a = 0\,,$$

whence

$$u + v\alpha + w\alpha^2 = -\frac{x_1^2 - a - 2\alpha x_1 - 2\alpha^2}{2y_1}$$

for $\alpha = \alpha_i$ and $i = 1, 2, 3$. This combined with the formula of Lemma 3 shows that $2(x_1, y_1) = (x_2, y_2)$, and Lemma 2 is proved.

The next step needs only the fact that K is the quotient field of a ring R which has unique factorization into prime elements.

Lemma 4. *Assume that $\alpha_i - \alpha_j$ is a unit in R for all $i \neq j$. Then for any $(x, y) \in A_K$ such that $x \neq \alpha_i$ for all i we have*

$$x - \alpha_i = t_i^2 u_i$$

for some element $t_i \in K$ and a unit u_i in R.

Proof. The product $(x - \alpha_1)(x - \alpha_2)(x - \alpha_3)$ is a square in K, and the factors $x - \alpha_i$ are relatively prime by assumption. The lemma is then obvious.

It follows from the lemma that the image of θ_i is contained in cosets of K^{*2}

represented by units of R. If K is a number field, its ring of algebraic integers $\mathfrak{o}_K$ is of course not a unique factorization domain. However, by localizing this ring at a finite number of primes (say those primes dividing ideals representing the finite number of ideal classes and those dividing $\alpha_i - \alpha_j$), one obtains a ring R which has unique factorization, and such that the unit group R is finitely generated. Therefore the images of the homomorphisms θ_i are contained in finite groups as we wished.

§ 2. The Kummer Pairing for Elliptic Curves

Throughout this section, we fix an integer $m \geqslant 2$, and we assume that A is an elliptic curve defined over K, such that $A_m \subset A_K$.

As usual, we take K to be a number field, but what we say would hold more generally if the characteristic of K does not divide m.

Let $Q \in A_K$. Any point P such that $mP = Q$ can be denoted by $\frac{1}{m}Q$. If P' is another such point, then

$$P' = P + t,$$

with some $t \in A_m$, and conversely. Of course, the point P need not be in A_K, its coordinates are algebraic over K. Since we have assumed that $A_m \subset A_K$, however, it follows that the extension

$$K(P) = K(P + t)$$

is well defined, independent of the choice of P such that $mP = Q$. We may therefore denote it by $K\left(\frac{1}{m}Q\right)$.

Observe that we have here a situation analogous to that of the ordinary m-th roots when dealing with the multiplicative group of K. Let $\alpha \in K^*$. If the m-th roots of unity are not contained in K, then the extension $K(\alpha^{1/m})$ is not well defined, since an m-th root of α is only defined up to multiplication by an m-th root of unity. If the m-th roots of unity are contained in K, then this extension is well defined, and is cyclic.

Returning to the elliptic curve, let σ be an isomorphism of $K\left(\frac{1}{m}Q\right)$ over K. If $mP = Q$ then $\sigma(mP) = m(\sigma P) = Q$, so

$$\sigma P = P + t_\sigma, \qquad \text{some } t_\sigma \in A_m.$$

Therefore σ induces an automorphism of $K\left(\frac{1}{m}Q\right)$, and the map

$$\sigma \mapsto t_\sigma$$

is a homomorphism

$$\mathrm{Gal}\left(K\left(\frac{1}{m}Q\right)/K\right) \to A_m .$$

This homomorphism is injective, because the effect of σ is determined by its effect on P (i.e. on the coordinates of P). Hence in particular, the Galois group is abelian, and isomorphic to a group of translations by elements of A_m.

More generally, let $K_m = K\left(\frac{1}{m}A_K\right)$ be the field obtained as the compositum of all fields $K\left(\frac{1}{m}Q\right)$ for $Q \in A_K$. Then K_m is abelian over K. Let G be its Galois group. There is a pairing

$$(G, A_K) \to A_m$$

obtained as follows. Let $Q \in A_K$. Let P be a point such that

$$mP = Q .$$

Let $\sigma \in G$. Let $t(\sigma, Q) \in A_m$ be such that $\sigma P = P + t(\sigma, Q)$. Then it is immediately verified that $t(\sigma, Q)$ is independent of the choice of P, and that the association

$$(\sigma, Q) \mapsto t(\sigma, Q)$$

is bilinear. If σ is orthogonal to A_K, then $\sigma = \mathrm{id}$ because σ leaves every generator P of K_m fixed. If Q is orthogonal to G, then a point P such that $mP = Q$ is fixed by G, and therefore is rational over K, that is $P \in A_K$. Hence $Q \in mA_K$. It is therefore clear that the kernel of the bilinear map in A_K is precisely mA_K, and we obtain a pairing

$$G \times (A_K/mA_K) \to A_m$$

whose kernels on each side are 1 and 0 respectively. This is the usual set up of Kummer theory.

As an application we find the criterion:

G is finite if and only if A_K/mA_K is finite.

In the next section, we shall indicate a general proof showing that G is finite. This will conclude our second proof of the weak Mordell–Weil Theorem.

We note that the homomorphisms θ_i in § 1 merely make the Kummer theory map explicit when $m = 2$.

§ 3. Second Proof of the Weak Mordell–Weil Theorem

Theorem 3.1. *Let K be a number field, and A an elliptic curve defined over K. Let m be a positive integer, and assume that $A_m \subset A_K$. Then $K\left(\frac{1}{m} A_K\right)$ is finite over K.*

Proof. Suppose that A is defined by the equation

$$y^2 = x^3 + ax + b, \qquad a, b \in \mathfrak{o}_K .$$

Let $\mathfrak{o} = \mathfrak{o}_K$. For almost all prime ideals $\mathfrak{p}$ of $\mathfrak{o}$ (all but a finite number) we can define an elliptic curve $\bar{A} = A \bmod \mathfrak{p}$, by the equation

$$y^2 = x^3 + \bar{a}x + \bar{b},$$

where $\bar{a}$ and $\bar{b}$ are $a \bmod \mathfrak{p}$ and $b \bmod \mathfrak{p}$ respectively. The algebraic formulas for addition and division of points can then be reduced mod $\mathfrak{p}$, and give addition and division on $\bar{A}$. Let $F_m(X)$ be the monic polynomial whose roots are the x-coordinates of the points in A_m. Because the group law on $\bar{A}$ is obtained by reduction mod $\mathfrak{p}$ of the group law on A, it follows that the points $\bar{P}$ in $\bar{A}_m$ are such that their x-coordinates are also the roots of the reduced equation

$$\bar{F}_m(X) = 0 .$$

If $F_m(X) = \sum \alpha_j X^j$ with $\alpha_j \in K$, then by definition,

$$\bar{F}_m(X) = \sum \bar{\alpha}_j X^j .$$

For almost all prime ideals $\mathfrak{p}$, the polynomial $\bar{F}_m$ has the same degree as F_m, and has distinct roots.

This gives rise to an injection

$$x(P) \mapsto \overline{x(P)}$$

on the x-coordinates of points of order m, $P \neq O$, whence reduction mod $\mathfrak{p}$

$$P \to \bar{P}$$

induces an isomorphism $A_m \to \bar{A}_m$. This holds for all $\mathfrak{p}$ outside a finite set of primes S. More precisely, S can be taken to consist of those primes where A has bad reduction, or dividing m.

Let now $Q \in A_K$. Let

$$E = K\left(\frac{1}{m} Q\right) = K(P),$$

where P is any point such that $mP = Q$. We know that the Galois group operates by translations, i.e. if σ is an element of the Galois group, then

$$\sigma P = P + t_\sigma$$

for some $t_\sigma \in A_m$. Let $\mathfrak{P}$ be a prime ideal in E extending some prime ideal $\mathfrak{p} \notin S$. We denote reduction mod $\mathfrak{P}$ also by putting a bar over elements in E. Then we find

$$\overline{\sigma P} = \overline{P} + \overline{t_\sigma}\,.$$

Since $A_m \to \overline{A}_m$ is an isomorphism, it follows that the induced effect of σ on the residue class field (if σ is in the decomposition group of $\mathfrak{P}$) cannot be the identity unless $\sigma = \text{id}$. Therefore $\mathfrak{P}$ is unramified over $\mathfrak{p}$.

This implies that $K\left(\frac{1}{m} A_K\right)$ is an abelian extension of K, which is unramified outside the finite set of primes in S, and such that every element of the Galois group has period dividing m. The ordinary Kummer theory shows that this extension can then be obtained by extracting m-th roots of elements $\alpha \in K$ which are of the following types:

α is a unit in the localized ring $\mathfrak{o}_S$, obtained by localizing $\mathfrak{o}_K$ at all primes in S.
α has an ideal factorization which is an m-th power.

Since the units in $\mathfrak{o}_S$ form a finitely generated group, and since the class number is finite, it follows that E is finite over K. This concludes the proof.

Corollary. *A_K/mA_K is finitely generated.*

Proof. Immediate from Theorem 3.1 and the Kummer theory of the preceding section.

Observe that the proof gives an effective bound for the order of the Galois group G in terms of the rank of the unit group $\mathfrak{o}_S^*$ and the S-class number. For simplicity, suppose that the set S is chosen so large that the S-class number is 1. Then the maximal abelian extension of K of exponent m, unramified outside S is obtained as

$$K(U^{1/m})$$

where U is the group of units. The Galois group of this extension is dual by Kummer theory to U/U^m, and in particular has order m^s where s is the number of elements in S.

Since we have an exact pairing

$$G \times A_K/mA_K \to A_m,$$

and therefore an injection

$$0 \to A_K/mA_K \to \mathrm{Hom}\,(G, A_m)\,,$$

and since $(A_K : mA_K) = m^{r+2}$ where r is the rank of A_K, we get a bound for the rank in terms of the rank of the units, namely

$$m^{r+2} \leqslant (\mathrm{card}\ G)^2 = m^{2s}$$

in other words

$$r \leqslant 2s - 2\,.$$

This bound is quite effective. What is not effective is to get a bound on the heights of generators of cosets of A_K/mA_K.

§ 4. Kummer Theory for the Multiplicative Group

Before dealing with more precise versions of Kummer theory on elliptic curves, we deal with the analogous situation on the multiplicative group.

We let K be a number field (even though everything we say here would hold much more generally in the obvious manner). We let Γ be a finitely generated multiplicative subgroup of K^*. We let

$$\begin{aligned} \Gamma' &= \textbf{division group of } \Gamma \text{ in } K^* \\ &= \{\beta \in K^*,\ \beta^m \in \Gamma \text{ for some positive integer } m\}\,. \end{aligned}$$

If we deal with an extension E of K, we write Γ'_E for the division group of Γ in E^*.

The index $(\Gamma' : \Gamma)$ is finite.

Let n be a positive integer prime to the index $(\Gamma' : \Gamma)$. Then we have isomorphisms

$$\boxed{\Gamma/\Gamma^n \approx \Gamma/\Gamma \cap K^{*n} \approx \Gamma K^{*n}/K^{*n}\,.}$$

This is obvious, because under the stated hypothesis,

$$\Gamma^n = \Gamma \cap K^{*n}\,.$$

In particular, for Γ' itself, we have

$$\boxed{\Gamma'/\Gamma'^n = \Gamma'/(\Gamma' \cap K^{*n}) \approx \Gamma' K^{*n}/K^{*n}\,.}$$

Ordinary Kummer theory of elementary algebra shows:

If the n-th roots of unity are contained in K, then

$$[K(\Gamma^{1/n}) : K] = n^{r+1}\,.$$

Cf. for instance my *Algebra*, Chapter VIII, § 8.

Let μ_n be the group of n-th root of unity. We are interested in the Galois extension

$$K_{n,\Gamma} = K(\mu_n, \Gamma^{1/n})$$

obtained by adjoining the n-th roots of unity and all n-th roots of elements in Γ. Let

$$E_n = K(\mu_n)\,.$$

Then we have a tower of fields

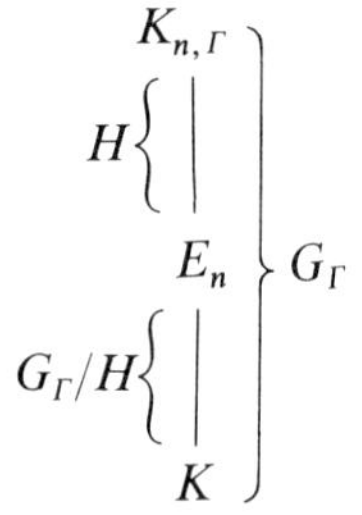

$$\left.\begin{array}{rc} & K_{n,\Gamma} \\ H\Big\{ & \Big| \\ & E_n \\ G_\Gamma/H\Big\{ & \Big| \\ & K \end{array}\right\} G_\Gamma$$

with groups as indicated:

$$H = \mathrm{Gal}\,(K_{n,\Gamma}/E_n) \quad \text{and} \quad G_\Gamma = \mathrm{Gal}\,(K_{n,\Gamma}/K).$$

The Kummer theory over E_n yields a natural pairing

$$H \times \Gamma/\Gamma^n \to \mu_n$$

which can be expressed as a homomorphism

$$\boxed{\Gamma/\Gamma^n \to \mathrm{Hom}\,(H, \mu_n)\,.}$$

To each element $\alpha \in \Gamma$ we associated the homomorphism

$$\varphi_\alpha \colon H \to \mu_n$$

such that

$$\varphi_\alpha(\tau) = \frac{\tau\alpha^{1/n}}{\alpha^{1/n}}\cdot$$

Here, $\alpha^{1/n}$ denotes any n-th root of α. The above quotient is independent of the chosen root of α, and is an n-th root of unity.

Theorem 4.1. *Assume:*

(i) *n is prime to* $2(\Gamma' : \Gamma)$.

(ii) $\mathrm{Gal}\,(K(\mu_n)/K) \approx (\mathbf{Z}/n\mathbf{Z})^*$.

Then the map

$$\Gamma/\Gamma^n \to \mathrm{Hom}\,(H, \mu_n)$$

is an isomorphism.

Proof. The ordinary abelian Kummer theory over the field $K(\mu_n)$ gives an isomorphism

$$\Gamma/(\Gamma \cap E_n^{*n}) \to \mathrm{Hom}\,(H, \mu_n)\,.$$

The whole problem is to show that

$$\Gamma \cap E_n^{*n} = \Gamma^n\,.$$

If these two groups are not equal, then for some prime $p \mid n$ there exists an element $\alpha \in \Gamma$ such that

$$\alpha = \beta^p \quad \text{with } \beta \in K(\mu_n) \quad \text{but} \quad \beta \notin K\,.$$

In other words, α is not a p-th power in K but becomes a p-th power in $K(\mu_n)$. The equation

$$X^p - \alpha = 0$$

is irreducible over K, and β has degree p over K. Hence it remains of degree p over

$$K(\mu_p)$$

which has degree $p - 1$. It is trivially verified that the Galois extension

$$K(\mu_p, \alpha^{1/p}) \text{ over } K$$

is not abelian. Since $K(\mu_n)$ is abelian over K, it follows that $\alpha^{1/p}$ has degree p over $K(\mu_n)$, a contradiction which proves the theorem.

We may give a non-canonical form to the theorem. Let

$$\{\alpha_1, \ldots, \alpha_r\}$$

be a basis for Γ modulo its torsion group, if Γ has rank r. Each α_j gives rise to a homomorphism

$$\varphi_j = \varphi_{\alpha_j}: H \to \mu_n ,$$

essentially identifying the effect of H on $\alpha_j^{1/n}$ by multiplicative translations with n-th roots of unity.

Theorem 4.2. *Under the hypotheses of the theorem, the map*

$$\varphi^{(r)}: H \to \mu_n^{(r)}$$

given by

$$\tau \mapsto (\varphi_1(\tau), \dots, \varphi_r(\tau))$$

is an isomorphism.

Proof. Abbreviate $\varphi^{(r)}$ by φ. Suppose $\varphi(\tau) = 1$. Then τ must be the identity on $E_n(\alpha_1^{1/n}, \dots, \alpha_r^{1/n})$, whence $\tau = 1$. Hence φ is injective. Theorem 4.1 implies that H has order n^r, which is also the order of $\mu_n^{(r)}$. Hence φ is surjective, and the theorem is proved.

Remark. By Theorem 5.1 of the preceding chapter, we see that the hypothesis of the theorem are satisfied if every prime dividing n satisfies

$$p \geqslant C_0 U^r$$

where C_0 is a sufficiently large constant depending only on r and $[K : \mathbf{Q}]$, and where

$$U = \max \log H_K(\alpha_j) .$$

This gives an explicit determination of those n for which the theorem applies.

Identifying $H \approx \mu_n^{(r)}$ in a situation like that of the theorem, we then have an exact sequence

$$1 \to H \approx \mu_n^{(r)} \to G_\Gamma \to GL_1(\mathbf{Z}/n\mathbf{Z}) \to 1 .$$

We may also pass to the limit, over integers n ordered by divisibility. It is easy to deal with the finite number of exceptional primes not satisfying the conditions of the theorem. We then obtain the following situation.

Let

$$K(\Gamma^{1/\infty}) = K_{\infty,\Gamma} = K_\infty$$

be the field obtained by adjoining all roots of unity and all division values of elements in Γ. Let

$$E_\infty = K(\mu)$$

be the field obtained by adjoining all roots of unity to K. We have the inclusion tower

$$\begin{array}{ccl} & K_\infty & \\ H_\infty\left\{\right. & | & \\ & E_\infty & \quad G_{\infty,\Gamma} = \mathrm{Gal}\,(K_\infty/K)\,. \\ (G_\Gamma/H)_\infty\left\{\right. & | & \\ & K & \end{array}$$

There is an exact sequence

$$1 \to H_\infty \to G_{\infty,\Gamma} \to (G_\Gamma/H)_\infty \to 1$$

and one can identify:

$$H_\infty \approx \text{open subgroup of } \prod \mathbf{Z}_p^{(r)}$$

$$(G_\Gamma/H)_\infty \approx \text{open subgroup of } \prod \mathbf{Z}_p^* = \prod GL_1(\mathbf{Z}_p)\,.$$

The proof consists in combining the theorem proved above for all but a finite number of primes, with a proof that for a finite number of primes, the group H_p is of finite index in $\mathbf{Z}_p^{(r)}$, and G_p is of finite index in $\mathbf{Z}_p^* = GL_1(\mathbf{Z}_p)$. One can use an irreducibility criterion for $X^{p^n} - a = 0$ as in my *Algebra*, Chapter VIII, § 9 (needed only in the simplest cases), or one can use a little cohomology as will be done for elliptic curves in the next section, where all the details will be carried out in an essentially harder case. Thus we leave the rest of the present case to the reader.

The local isogeny theorem. As mentioned before, the Kummer theory also holds over arbitrary fields of characteristic not dividing n. As an application, we give a simplified proof of a theorem of Serre, cf. [L 2] Chapter 16, § 3.

Local isogeny theorem. *Let A, B be elliptic curves defined over a finite extension of $\mathbf{Q}_p$, with invariants j, j' such that* $\mathrm{ord}_p j$ *and* $\mathrm{ord}_p j' < 0$. *Assume that*

$$K(A^{(p)}) = K(B^{(p)})\,.$$

Then A and B are isogenous.

Proof. We assume that the reader knows the p-adic analytic isomorphism of the curves with the multiplicative group, i.e. that A, B are Tate curves possibly over a finite extension K. Cf. [L 2], Chapter 15, § 1. The invariants j, j' correspond to parameters q, q' in K, and after taking powers if necessary, we may assume that q, q' have the same order at p, so q'/q is a unit u. We need to show that u is a root of unity. Making a finite extension if necessary, we may assume that K contains the p^2 roots of unity (p-th roots of unity would be sufficient if $p \neq 2$). Let

$$\Gamma = \{\zeta, q, u\}$$

be the group generated by a primitive p-th root of unity, q and u. The next theorem gives a contradiction if u is not a root of unity.

Theorem 4.3. *Let Γ be a finitely generated multiplicative subgroup of a p-adic field K. Let Γ' be its p-division group in K^*. Assume that the p^2-th roots of unity are in K. Let*

$$(\Gamma' : \Gamma'^p) = p^r .$$

Then

$$[K(\Gamma'^{1/p^n}) : K] = p^{rn} .$$

Proof. We prove in fact that $[K(\Gamma'^{1/p}) : K] = p^r$ and that $\Gamma'^{1/p}$ is its own p-division group in $K(\Gamma'^{1/p})$. Since

$$\Gamma' K^{*p}/K^{*p} \approx \Gamma'/\Gamma' \cap K^{*p} = \Gamma'/\Gamma'^p$$

has rank r over $\mathbf{Z}/p\mathbf{Z}$, we see by ordinary Kummer theory that the degree of the stated extension is p^r. Let $v \in K(\Gamma'^{1/p})$, and $v^p \in \Gamma'^{1/p}$. Then $v^{p^2} = w \in \Gamma'$. Then w is a p-th power in Γ', for otherwise

$$X^{p^2} - w = 0$$

is irreducible, and the Galois group of $K(\Gamma'^{1/p})$ contains an element of exact order p^2, which is impossible. Then

$$(v^p)^p = u^p \quad \text{with some } u \text{ in } \Gamma' ,$$

whence v^p is in Γ', as was to be shown. Also note that under the p-th power we have an isomorphism

$$\Gamma'^{1/p}/\Gamma' \approx \Gamma'/\Gamma'^p ,$$

so the rank condition is preserved. This proves the theorem.

Remark. The method also provides a generalized local isogeny theorem for several elliptic curves, not just two of them. What we show is essentially that if elements are multiplicatively dependent, then the Kummer extensions which they generate are also independent. The Tate analytic isomorphism allows us to transport this to elliptic curves with j-invariants which are not integral at p.

§ 5. Bashmakov's Theorem

In this section, we carry out for elliptic curves the analogue of the last section for the multiplicative group. We let A be the usual elliptic curve defined over the number field K. We denote the group of rational points over K by $A(K)$, the group of n-torsion points by A_n. We let Γ be a given finitely generated subgroup of $A(K)$. We let:

$$E_n = K(A_n)$$

$$K_{n,\Gamma} = K_n = K\left(A_n, \frac{1}{n}\Gamma\right).$$

Thus K_n is obtained from K by adjoining all coordinates of n-torsion points, and n-division points from elements of Γ. We have a tower of fields with associated Galois groups:

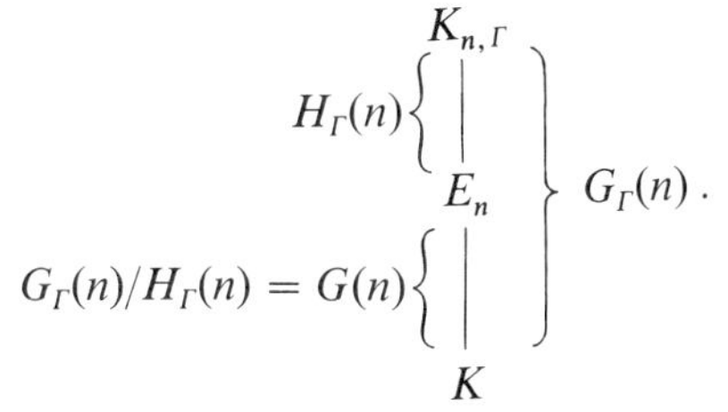

$$\begin{array}{ccc} & K_{n,\Gamma} & \\ H_\Gamma(n)\left\{\right. & \Big| & \\ & E_n & \Bigg\} \; G_\Gamma(n)\,. \\ G_\Gamma(n)/H_\Gamma(n) = G(n)\left\{\right. & \Big| & \\ & K & \end{array}$$

We often omit n from the notation, writing H_Γ or even H instead of $H_\Gamma(n)$, and similarly G_Γ instead of $G_\Gamma(n)$. We also write

$$G = G(n) = G_\Gamma(n)/H_\Gamma(n)\,.$$

The Kummer theory over E_n yields a pairing

$$H_\Gamma \times \Gamma/(\Gamma \cap nA(E_n)) \to A_n\,.$$

This is an exact pairing, i.e. the kernels on each side on the left are reduced to the unit elements of each group. We therefore get a pairing

$$H_\Gamma \times \Gamma/n\Gamma \to A_n$$

which can be expressed as a homomorphism

$$\varphi\colon \Gamma/n\Gamma \to \mathrm{Hom}\,(H_\Gamma, A_n)\,.$$

To each $P \in \Gamma$ we associated the homomorphism

$$\varphi_P \colon H_\Gamma \to A_n$$

such that

$$\varphi_P(\tau) = \tau Q - Q\,,$$

where Q is any point such that $nQ = P$. The above difference is independent of the chosen "n-th root" Q of P, and is an n-torsion point.

Since H is a normal subgroup of G_Γ, the group G_Γ (actually G_Γ/H) operates on H by conjugation,

$$\tau \mapsto \sigma\tau\sigma^{-1}\,.$$

It is immediately verified that φ_P is a G_Γ/H-homomorphism, i.e.

$$\boxed{\varphi_P(\sigma\tau\sigma^{-1}) = \sigma\varphi_P(\tau)\,.}$$

Therefore our map φ is actually an arrow

$$\varphi \colon \Gamma/n\Gamma \to \mathrm{Hom}_G\,(H, A_n)\,.$$

Suppose furthermore that γ is an endomorphism of A, and that γ is defined over K. Then the association $P \mapsto \varphi_P$ commutes with γ, namely

$$\varphi_{\gamma P}(\tau) = \gamma\varphi_P(\tau)\,, \qquad \text{all } \tau \in H.$$

Indeed, if $P = nQ$, then $\gamma P = n\gamma Q$, and therefore $\tau Q - Q = t$ implies

$$\tau(\gamma Q) - \gamma Q = \gamma t = \gamma\varphi_P(\tau)\,.$$

This proves the desired commutation.

We shall need a lemma applicable in general, showing that non-degeneracy at arbitrary levels can be reduced to non-degeneracy at prime level.

For any positive integer n we denote by

$$A^{(n)}$$

the group of torsion points on A whose order divides a power of n. In particular, $A^{(l)}$ is the group of l-power torsion points on A.

Let E be a field over which all points of A_{l^m} are rational. We allow $m = \infty$, in which case all points of $A^{(l)}$ are rational over E. We denote

$$H_\Gamma(l^m, E) = \mathrm{Gal}\left(E\left(\frac{1}{l^m}\Gamma\right)\Big/E\right).$$

We assume that Γ is free of rank r over $\mathfrak{o}$. Let $\{P_1, \ldots, P_r\}$ be a basis of Γ over $\mathfrak{o}$ $= \operatorname{End} A$. We have a map

$$\tau \mapsto (\varphi_{P_1}(\tau), \ldots, \varphi_{P_r}(\tau))$$

of

$$H(l^m, E) \to A_{l^m}^{(r)} .$$

It will also be convenient to pass to the projective limit. We recall that

$T_l = T_l(A)$ is the group of vectors

$$(a_1, a_2, \ldots)$$

such that $a_m \in A_{l^m}$ and $la_{m+1} = a_m$. To each point P_j we introduce an infinite vector

$$Y_j = (Q_{j,0}, Q_{j,1}, Q_{j,2}, \ldots)$$

where

$$Q_{j,0} = P_j \quad \text{and} \quad lQ_{j,m+1} = Q_{j,m} .$$

For each $\tau \in H(l^\infty, E)$ we have an element

$$\varphi_j(\tau) = \tau Y_j - Y_j \in T_l(A) ,$$

and the map

$$\varphi : \tau \mapsto (\varphi_1(\tau), \ldots, \varphi_r(\tau))$$

gives a homomorphism

$$H(l^\infty, E) \to T_l(A)^{(r)} .$$

Lemma 1. *If the map*

$$H(l, E) \to A_l^{(r)}$$

is an isomorphism, then the map

$$H(l^\infty, E) \to T_l^{(r)}$$

is an isomorphism, and for each m,

$$H(l^m, E) \to A_{l^m}^{(r)}$$

is an isomorphism.

Proof. We have a commutative diagram

$$\begin{array}{ccc} H(l^\infty, E) & \longrightarrow & T_l^{(r)} \\ \downarrow & & \downarrow \mathrm{pr}_1 \\ H(l) & \xrightarrow{\approx} & A_l^{(r)} \end{array}$$

The right arrow is projection on the first coordinate. The left arrow is surjective, and the bottom arrow is an isomorphism by hypothesis. Hence pr_1 is surjective. The proof that the top arrow is an isomorphism (i.e. surjective) then results from the following lemma.

Lemma 2. *Let W be a closed subgroup of $T_l^{(r)}$ whose projection on $A_l^{(r)}$ is surjective. Then $W = T_l^{(r)}$.*

Proof. This is a simple exercise. If $\xi_1, \ldots, \xi_{2r}$ are elements of W whose projections on the first component form a basis of $A_l^{(r)}$ over $\mathbf{Z}/l\mathbf{Z}$, then an easy refinement procedure shows that they form a basis of $T_l^{(r)}$ over $\mathbf{Z}_l$. We can leave this to the reader, who can also look it up for instance in [L 2], Chapter 13, § 1, proof of Theorem 1 (replace 1 by r). The whole thing can also be viewed as a special case of Nakayama's lemma, cf. *Algebra*, p. 242, or 155.

We also need simple notions of cohomology.

Let G be a group and M a G-module. A function

$$f\colon G \to M$$

is called a **1-cocycle** if

$$f(\sigma) + \sigma f(\tau) = f(\sigma\tau)$$

for all $\sigma, \tau \in G$. The cocycles form a group. If $a \in M$, then the function

$$\sigma \mapsto \sigma a - a$$

is a cocycle, called a **coboundary**. The coboundaries form a group. The factor group is denoted by

$$H^1(G, M),$$

and is called the first **cohomology group**.

Some of the results of Bashmakov on cohomology are contained in the following general theorem of Sah [Sa].

Theorem 5.1. *Let G be a group and let M be a G-module. Let α be in the center of G. Then $H^1(G, M)$ is annihilated by the map $x \mapsto \alpha x - x$ on M. In particular, if this map is an automorphism of M, then $H^1(G, M) = 0$.*

Proof. The theorem is in fact valid for H^r (see below). The following direct proof for H^1 was shown to me by W. Ellis. First we note that

$$f(1) = f(1 \cdot 1) = f(1) + 1 \cdot f(1) = 2f(1),$$

so that $f(1) = 0$. Also,

$$0 = f(1) = f(\alpha\alpha^{-1}) = f(\alpha) + \alpha f(\alpha^{-1}).$$

Hence for arbitrary σ, we have

$$\begin{aligned} f(\sigma) = f(\alpha\sigma\alpha^{-1}) &= f(\alpha) + \alpha f(\sigma\alpha^{-1}) \\ &= f(\alpha) + \alpha[f(\sigma) + \sigma f(\alpha^{-1})]. \end{aligned}$$

Therefore

$$\alpha f(\sigma) - f(\sigma) = f(\alpha) + \sigma\alpha f(\alpha^{-1}),$$

which proves the theorem.

Using a little more cohomology of groups, one obtains the general version of Sah. Let $H^r(\alpha)$ be the induced homomorphism on the cohomology. It is a standard fact that $H^r(\alpha) = I$. However, from the standard complex or otherwise, one sees that $H^r(\alpha)$ is the same map as that induced by

$$\alpha\colon M \mapsto M, \qquad x \mapsto \alpha x,$$

for instance because $H^r(\alpha)$ on a standard cochain is given by

$$\alpha f(\alpha^{-1}\sigma_1\alpha, \ldots, \alpha^{-1}\sigma_r\alpha) = \alpha f(\sigma_1, \ldots, \sigma_r).$$

Hence $H^r(\alpha) - I = 0$. This proves the result.

In the applications, suppose $G = GL_d(R)$ where R is some ring. To apply the theorem, all we need is that the group of units R^* contains an element α such that $\alpha - 1$ is also a unit. If R is $\mathbf{Z}/n\mathbf{Z}$ for some positive integer n, then R has this property if n is not divisible by 2. If R is a finite field with >2 elements, then R has this property.

At some point of the proof, we have to deal separately with the complex multiplication case and the non-CM case. We make some preliminary remarks on complex multiplication for this purpose. We assume that the reader is acquainted with the theory of complex multiplication. He can find expositions in Shimura's book [Sh 1] or Lang [L 2].

We let $\mathfrak{o} = \text{End}\, A$ be the ring of endomorphisms of A. Suppose that the quotient field of $\mathfrak{o}$ is an imaginary quadratic field k. In [L 2] Chapter VIII, § 1 will be found a discussion of the structure of $\mathfrak{o}$. In particular, one sees that if n is a positive integer prime to the conductor of $\mathfrak{o}$, then

$$\mathfrak{o}/n\mathfrak{o} \approx \mathfrak{o}_k/n\mathfrak{o}_k$$

where $\mathfrak{o}_k$ is the ring of algebraic integers in k. We abbreviate

$$\mathfrak{o}(n) = \mathfrak{o}/n\mathfrak{o} .$$

Under the exponential map there is a natural isomorphism

$$\mathbf{C}/\mathfrak{a} \approx A_{\mathbf{C}}$$

where $\mathfrak{a}$ is a proper $\mathfrak{o}$-ideal, and therefore an isomorphism

$$k/\mathfrak{a} \approx A_{\text{tor}} ,$$

whence an isomorphism

$$\frac{1}{n}\mathfrak{a}/\mathfrak{a} \approx A_n .$$

This is also an $\mathfrak{o}$-isomorphism. Since locally every ideal of $\mathfrak{o}_k$ is principal, we obtain:

Fact 1. *A_n is a* 1*-dimensional free* $\mathfrak{o}(n)$*-module.*

We shall assume that $k \subset K$. Then one has a simple model for the Galois group of $K(A_{\text{tor}})$ over K, which is necessarily abelian. This is essentially a classical result (Hasse–Deuring), which was put in idele form by Shimura, cf. [Sh], 7.8 and also [L 2], Chapter 10, § 4, Theorem 8. For each prime l let $\mathfrak{o}_l$ be the l-adic completion of $\mathfrak{o}$, and let $\mathfrak{o}_l^*$ be the group of units. Then there is an embedding

$$\text{Gal}\,(K(A_{\text{tor}})/K) \to \prod_l \mathfrak{o}_l^*$$

onto an open subgroup of the product. In particular:

Fact 2. *There is a finite set of primes S such that if n is prime to S, then*

$$G(n) = \text{Gal}\,(K(A_n)/K) \approx \mathfrak{o}(n)^* .$$

Note also that we can enlarge the set S so that if n is prime to S then

$$\mathfrak{o}(n) = \mathbf{Z}(n)[\mathfrak{o}(n)^*] ,$$

in other words, $\mathfrak{o}(n)$ is generated over $\mathbf{Z}(n)$ by the units in $\mathfrak{o}(n)$.

The next lemma will be used only when $n = l$ is prime, but we state it in general because of its intrinsic interest.

Lemma 3. *Let W be an $\mathfrak{o}(n)$-submodule of $\mathfrak{o}(n)^{(r)}$. If*

$$W \neq \mathfrak{o}(n)^{(r)},$$

then there exist elements $\alpha_1, \ldots, \alpha_r \in \mathfrak{o}(n)$ not all 0 such that for all $(w_1, \ldots, w_r) \in W$ we have

$$\alpha_1 w_1 + \cdots + \alpha_r w_r = 0 .$$

Proof. We have a duality

$$\mathfrak{o}(n)^{(r)} \times \mathfrak{o}(n)^{(r)} \to \mathfrak{o}(n)$$

given by the usual scalar product of r-tuples. In this way,

$$\mathfrak{o}(n)^{(r)} \approx \operatorname{Hom}_{\mathfrak{o}(n)}(\mathfrak{o}(n)^{(r)}, \mathfrak{o}(n)) .$$

On the other hand, $\mathfrak{o}(n)$ is the reduction mod n of a principal ring obtained from $\mathfrak{o}$ by localizing at all primes except those dividing n. We can apply the elementary divisor theory to the free module over this ring, and then reduce mod n again. Thus we can construct a basis

$$\{u_1, \ldots, u_r\}$$

of $\mathfrak{o}(n)^{(r)}$ such that a basis for W is given by $\{\gamma_1 u_1, \ldots, \gamma_r u_r\}$ with some elements $\gamma_1, \ldots, \gamma_r \in \mathfrak{o}(n)$ and some element γ_j is not a unit. We pick $\beta \in \mathfrak{o}(n)$ such that $\beta\gamma_j = 0$ but $\beta \neq 0$. Then the map λ obtained by projection on the u_j-coordinate, followed by multiplication with β, is $\mathfrak{o}(n)$-linear on $\mathfrak{o}(n)^{(r)}$ and vanishes on W. The r-tuple $(\alpha_1, \ldots, \alpha_r)$ associated with λ in the standard basis for $\mathfrak{o}(n)^{(r)}$ satisfies the required condition.

Suppose next that $\mathfrak{o} = \mathbf{Z}$, so that A does not have complex multiplication. In that case, we recall that Serre has proved [Se 2]:

Fact 3. *There exists a finite set of primes S such that if n is prime to S, then*

$$G(n) = \operatorname{Gal}(K(A_n)/K) \approx GL_2(\mathbf{Z}/n\mathbf{Z}) .$$

We also denote $GL_2(\mathbf{Z}/n\mathbf{Z})$ by $GL_2(n)$.

By Sah's theorem, we see that if n is not divisible by 2, then

$$H^1(G(n), A_l) = 0$$

whenever $G(n) = \mathfrak{o}(n)^*$ or $G(n) = GL_2(n)$.

If Γ is a subgroup of $A(K)$, we let

$$\begin{aligned} \Gamma' &= \text{division group of } \Gamma \text{ in } A(K) \\ &= \text{set of points } Q \in A(K) \text{ such that } mQ \in \Gamma \text{ for some integer } m \geqslant 1. \end{aligned}$$

Theorem 5.2. *Let Γ be a subgroup of $A(K)$, free of rank r over $\mathfrak{o} = \operatorname{End}(A)$. Let $n > 1$ be an integer such that:*

(0) *If A has a complex multiplication, then n is prime to the conductor of $\mathfrak{o}$, the discriminant of k, and*

$$\mathfrak{o}(n) = \mathbf{Z}(n)[\mathfrak{o}(n)^*] .$$

(i) *n is prime to $2(\Gamma':\Gamma)$.*

(ii) *$G(n) = \mathfrak{o}(n)^*$ in the complex multiplication case*
$G(n) = GL_2(n)$ in the non-CM case.

Let $\{P_1, \ldots, P_r\}$ be a basis of Γ over $\mathfrak{o}$. Then the map

$$\tau \mapsto (\varphi_{P_1}(\tau), \ldots, \varphi_{P_r}(\tau))$$

is an isomorphism

$$H_\Gamma(n) \xrightarrow{\approx} A_n^{(r)} .$$

Proof. We give the proof in steps. We use l to denote any prime dividing n. We let $P \in \Gamma$.

It will be convenient to use the notation

$$H(l^m, n) = \operatorname{Gal}\left(\mathrm{K}\left(A_n, \frac{1}{l^m}\Gamma\right)/K(A_n)\right).$$

We are especially interested in level l, lifted over the field $K(A_n)$.

$$\begin{array}{c} K\left(A_n, \frac{1}{l}\Gamma\right) \\ | \\ K(A_n) \\ | \\ K \end{array} \quad \left.\begin{array}{c} \\ \\ \\ \end{array}\right\} H(l, n) .$$

Step 1. *The map $\Gamma/l\Gamma \to \operatorname{Hom}(H(l,n), A_l)$ given by*

$$P \mapsto \varphi_P$$

is injective, and its image lies in $\mathrm{Hom}_{G(n)}(H(l, n), A_l)$.

Proof. Suppose φ_P is trivial. If $P = lQ$, then Q is rational over $K(A_n)$. The map

$$\sigma \mapsto \sigma Q - Q$$

is a 1-cocycle of $G(n)$ into A_l, and by Theorem 5.1 there exists an element $a \in A_l$ such that

$$\sigma Q - Q = \sigma a - a\,.$$

Then $Q + a$ is fixed under $G(n)$, whence $Q + a$ is rational over K. But $P = l(Q + a)$, whence P is l-th multiple in $A(K)$, whence in Γ, thereby proving Step 1.

Step 2. *Let* $\{P_1, \ldots, P_r\}$ *be a basis of* Γ. *The map*

$$\tau \mapsto (\varphi_{P_1}(\tau), \ldots, \varphi_{P_r}(\tau))$$

is an isomorphism

$$H(l, n) \approx A_l^{(r)}\,.$$

Proof. From the injection of Step 1, we conclude that

$$\mathrm{Hom}_{G(l)}(H(l, n), A_l)$$

has cardinality $\geqq l^r$. On the other hand, let W be the image of $H(l, n)$ under the map

$$\varphi : \tau \mapsto (\varphi_{P_1}(\tau), \ldots, \varphi_{P_r}(\tau))\,.$$

Complex multiplication case. The image W is an $\mathfrak{o}(n)$-submodule of $A_l^{(r)}$. If $W \neq A_l^{(r)} \approx \mathfrak{o}(l)^{(r)}$, then we apply Lemma 3 to get a relation

$$\alpha_1 \varphi_{P_1}(\tau) + \cdots + \alpha_r \varphi_{P_r}(\tau) = 0$$

for all $\tau \in H$. Let $P = \alpha_1 P_1 + \cdots + \alpha_r P_r$. Then $P \neq O$, and

$$\varphi_P(\tau) = 0$$

for all $\tau \in H$. This contradicts Step 3, and concludes the proof of this step.

Non-CM case. Now W is again a $G(l)$-submodule of $A_l^{(r)}$, and in the present case is semisimple. Since A_l is simple, this implies that W is $G(l)$-isomorphic to a direct sum of copies of A_l, and the number of elements in this sum must be r by Step 1. It follows that $W = A_l^{(r)}$, as desired.

We may now pass to the limit. We let

$$H(l^\infty, n^\infty) = \mathrm{Gal}\left(K\left(A^{(n)}, \frac{1}{l^\infty}\Gamma\right)/K(A^{(n)})\right),$$

where $A^{(n)}$ is the group of torsion points on A of order dividing some power of n.

Step 3. *The map*

$$\tau \mapsto (\varphi_{P_1}(\tau), \ldots, \varphi_{P_r}(\tau))$$

gives an isomorphism

$$H(l^\infty, n^\infty) \xrightarrow{\approx} T_l(A)^{(r)}.$$

Proof. This follows from Lemma 1 at the very beginning.

Step 4. *Write $n = l^m n'$ with $l \nmid n'$. Then the extensions*

$$K\left(A_n, \frac{1}{l^m}\Gamma\right) \quad \text{and} \quad K\left(A_n, \frac{1}{n'}\Gamma\right)$$

are disjoint over $K(A_n)$.

Proof. This is obvious because their degrees are relatively prime.

Theorem 5.2 is now immediate by induction on the number of prime factors of n.

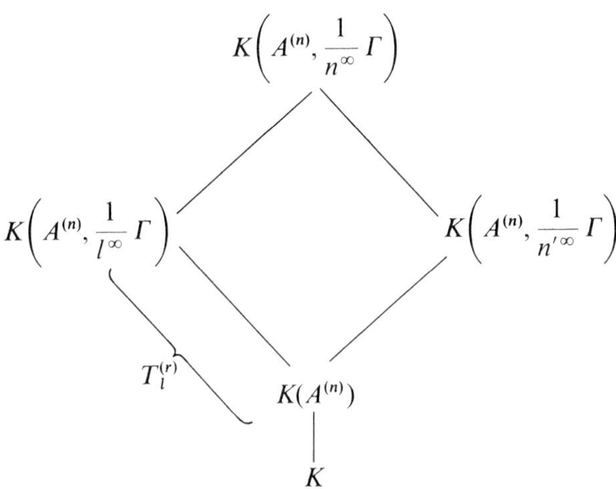

We can also express the theorem in its projective limit formulation. In the tower

$$G_\Gamma(n^\infty)\left\{\begin{array}{l} \left.\begin{array}{c} K\left(\dfrac{1}{n^\infty}\Gamma\right) \\ | \\ K(A^{(n)}) \end{array}\right\} H_\Gamma(n^\infty) \\ \left.\begin{array}{c} | \\ K \end{array}\right\} G(n^\infty) \end{array}\right.$$

we have the exact sequence

$$0 \to H_\Gamma(n^\infty) \to G_\Gamma(n^\infty) \to G(n^\infty) \to 0 .$$

Furthermore,

$$H_\Gamma(n^\infty) \approx T_n^{(r)}$$

$$G(n^\infty) \approx \mathfrak{o}_n^{(r)} .$$

We can let n grow by divisibility so as to include formal infinite products of primes for which the conditions of Theorem 5.2 are satisfied, and then the above remains true in the limit.

We shall now deal with the remaining cases of exceptional primes l when we do not get the full isomorphisms, but when $H_\Gamma(l^\infty)$ is only embedded as a subgroup of finite index in $T_l^{(r)}$.

There exists a positive integer M (divisible by 2) such that the Galois group of all torsion points G can be written

$$G = G_M \times \prod_{l \nmid M} G_l ,$$

G_M is open in $\prod_{l \mid M} G_l$, and $G_l = \mathfrak{o}_l^*$ or $GL_2(\mathbf{Z}_l)$ according as there is complex multiplication or not. We also write

$$G_{[M]} = \prod_{l \nmid M} G_l .$$

Each element $\sigma \in G$ has accordingly a product decomposition

$$\sigma = \sigma_M \times \sigma_{[M]} .$$

Selecting M to be sufficiently divisible, we can take an element in the center to be represented by a scalar matrix on each factor, namely $\alpha = \alpha_M \times \alpha_{[M]}$, where

$$\alpha_M = (1 + M)I_M \quad \text{and} \quad \alpha_{[M]} = 2I_{[M]},$$

using I to denote the unit 2×2 matrix.

Step 1. *Let $l \mid M$. Let $\lambda = M(\Gamma' : \Gamma)$. Let $P \in \Gamma$, and*

$$\varphi_P : H(l^m) \to A_{l^m}.$$

If $\varphi_P = 0$ then $\lambda P \in l^m \Gamma$.

Proof. Suppose $\varphi_P(\tau) = 0$ for all $\tau \in H(l^m)$. Then for any point Q such that $l^m Q = P$, we have

$$\varphi_P(\sigma) = \sigma Q - Q, \quad \sigma \in G.$$

By the cohomology lemma, there exists $a \in A_{l^m}$ such that

$$(\alpha - 1)(\sigma Q - Q) = \sigma a - a.$$

But $\sigma Q - Q \in A_{l^m}$, and G operates on A_{l^m} through G_M, so that the preceding equation can also be written in the form

$$M(\sigma Q - Q) = \sigma a - a = m(\sigma b - b)$$

for some point b such that $a = Mb$. Then $M(Q - b)$ is fixed under all $\sigma \in G$, that is $M(Q - b) \in A_K$, and

$$mP = Ml^m Q = l^m M(Q - b).$$

It follows that $M(\Gamma' : \Gamma)P \in l^m \Gamma$, as desired.

Over $K(A_{\text{tor}})$, we have the map

$$\varphi : \tau \mapsto (\varphi_{P_1}(\tau), \ldots, \varphi_{P_r}(\tau)) \in T_l(A)^{(r)}.$$

Then W is G_l-module, so a module over $\mathbf{Z}_l[G_l]$, generated by G_l over $\mathbf{Z}_l$.

CM case. Since G_l is isomorphic to a subgroup of $\mathfrak{o}_l^*$, of finite index, the ring $\mathbf{Z}_l[G_l]$ is a local order in $k_l = \mathbf{Q}_l \otimes k$. Furthermore, $k_l \otimes W$ is a k_l-submodule of $V_l(A)^{(r)}$, and is semisimple. If $k_l \otimes W \neq V_l(A)^{(r)}$, then there exist $\beta_1, \ldots, \beta_r$ in k_l not all 0 such that

$$\beta_1 \varphi_{P_1}(\tau) + \cdots + \beta_r \varphi_{P_r}(\tau) = 0$$

for all τ. We clear denominators, i.e. find a positive integer c such that $c\beta_j = \alpha_j \in \mathfrak{o}_l$ for $j = 1, \ldots, r$. Then

$$\alpha_1 \varphi_{P_1}(\tau) + \cdots + \alpha_r \varphi_{P_r}(\tau) = 0 \,,$$

and not all $\alpha_j = 0$. Let $\alpha_j^{(m)} \in \mathfrak{o}$ and $\alpha_j^{(m)} \equiv \alpha_j \bmod l^m$. Let

$$P = \alpha_1^{(m)} P_1 + \cdots + \alpha_r^{(m)} P_r \,.$$

Then $\varphi_P = 0$ on $H(l^m)$, whence by step 1,

$$\lambda P = \lambda \alpha_1^{(m)} P_1 + \cdots + \lambda \alpha_r^{(m)} P_r \in l^m \Gamma \,.$$

This is impossible for m sufficiently large, and hence

$$k_l \otimes W = V_l(A)^{(r)} \,.$$

It follows that W has finite index in $T_l(A)^{(r)}$, as desired.

No CM. We argue in the same way, replacing k_l by $\mathbf{Q}_l$, to see that $\varphi_{P_1}, \ldots, \varphi_{P_r}$ are linearly independent over $\mathbf{Z}_l$. On the other hand, in the present case, G_l is open in $GL_2(\mathbf{Z}_l)$, so $\mathrm{End}_{G_l}\, T_l(A) = \mathbf{Z}_l$ consists only of the scalars. Now $\mathbf{Q}_l \otimes W$ is a semisimple submodule of $V_l(A)^{(r)}$, and $\varphi_{P_1}, \ldots, \varphi_{P_r}$ extend to $\mathbf{Q}_l$-linearly independent elements of

$$\mathrm{Hom}_{G_l}(\mathbf{Q}_l \otimes W, V_l(A)) \,.$$

Since $V_l(A)$ is G_l-simple, it follows that $\mathbf{Q}_l \otimes W = V_l(A)^{(r)}$, whence as before W is open in $T_l(A)^{(r)}$, as desired.

Chapter VI. Integral Points

Siegel [Si 2] proved that on an arbitrary affine curve of genus $\geqslant 1$ there exist only a finite number of integral points. In dealing with hyperelliptic equations [Si 1] he uses a somewhat different principle. He first reduces the existence of integral points on the curve to the existence of solutions of another equation

$$(*) \qquad a\alpha + b\alpha' = 1 ,$$

where a, b are fixed coefficients, and α, α' are units in a number field. He then selects an integer $n \geqslant 3$, and coset representatives for the factor group U/U^n, where U is the group of units. Using these, the above equation becomes equivalent with the equation

$$a_1\alpha^n + b_1\alpha'^n = 1 ,$$

again to be solved in units α, α'. This makes a further reduction to a curve of higher genus. He then uses a theorem on diophantine approximations which ultimately became known as the Thue–Siegel–Roth theorem, to conclude the proof.

Taking another point of view in 1952, Gelfond pointed out in special cases, e.g. for the curves

$$f(x, y) = 1$$

where f is a binary form, that once equation (*) has been obtained, the finiteness of solutions follows from a lower bound for linear combinations of logarithms of algebraic numbers with integer coefficients, and even called the finding of such lower bounds one of the basic problems of diophantine approximations [Ge], last page of the book. He could prove such lower bounds only for linear combinations of two logarithms. Baker in 1966 saw how to prove them in general, thus completing the proof of finiteness of integral points by this approach, cf. [Ba 1] and the bibliography in that book.

In § 1 we shall describe the general technique along Gelfond's idea, which reduces the equation $a\alpha + b\alpha' = 1$ in units to a diophantine inequality following [L 5]. In § 2, § 3 we show how to reduce the analysis of integral points on elliptic curves to that equation in units, following Siegel [Si 1] and Chabauty [Ch]. In § 4 we give the reduction of an arbitrary model to the Weierstrass model. Only § 3, § 4 and § 7 use slightly more advanced knowledge about function fields in one variable and curves, and these may be omitted by a reader who wishes merely to see a proof as fast as possible for the finiteness of integral points on a given Weierstrass model.

There exists a fairly wide class of curves for which a direct reduction to the unit equation above can be made, because their affine rings of rational functions contain two units u and u' such that

$$u + u' = 1 .$$

For this, see Kubert–Lang [KL].

Baker's theorem also gave impetus to a reconsideration of the effectivity and efficiency of the unit theorem, as in Siegel [Si 2]. For applications to the estimate of integral points, cf. also Stark [St 2].

On the other hand, in 1964 I pointed out that instead of reducing the analysis of integral points to that equation, one could argue directly on the elliptic curve, and obtain the bound for integral points directly from a diophantine inequality concerning elliptic logarithms [L 5]. We shall reproduce the argument in § 8. The first inequality of this type, strong enough to suffice for the proof of the finiteness of integral points was given by Masser [Mas 1]. In a later chapter, we shall give the stronger inequality proved subsequently by Coates–Lang [Co–La]. All these inequalities on elliptic curves have been proved so far only when the elliptic curve has complex multiplication, and remain open in the general case when only a few special results are known involving one or two logarithms, cf. Baker [Ba 5], Coates [Co 3], [Co 4], Feldman [Fe 1], [Fe 2], Masser [Mas 1].

§ 1. The Equation $a\alpha + b\alpha' = 1$ in Units

Let K be a number field, and let $a, b \in K$. We are interested in the solutions of the equation

$$a\alpha + b\alpha' = 1$$

with units α, α' in $\mathfrak{o}_K$, or more generally in a finitely generated multiplicative subgroup Γ of K. There exists a finite set S of absolute values in M_K, containing all the archimedean absolute values S_∞, such that Γ is contained in the S-units K_S, i.e. those elements of K^* having absolute value equal to 1 outside S. Let $s + 1$ be the cardinality of S, and map

$$L: K_S \to \mathbf{R}^{s+1}$$

by the log, such that

$$L(\alpha) = (v_1(\alpha), \ldots, v_{s+1}(\alpha)) ,$$

where v_j are the elements of S, and we define for convenience

$$v(\alpha) = -\log \|\alpha\|_v, \qquad \|\alpha\|_v = |\alpha|_v^{n_v}.$$

The kernel of L consists precisely of the roots of unity in K. The image of L is a lattice of dimension s in the hyperplane $\mathbf{H}$ defined by the equation

$$\sum_{j=1}^{s+1} x_j = 0.$$

We define a function $\mathbf{F}$ on $\mathbf{R}^{s+1}$ by the condition

$$\mathbf{F}(X) = \sum_{j=1}^{s+1} \sup(0, x_j).$$

Then for any $t \geqslant 0$ we have $\mathbf{F}(tX) = t\mathbf{F}(X)$, and for any two vectors X, Y we have $\mathbf{F}(X + Y) \leqslant \mathbf{F}(X) + \mathbf{F}(Y)$. We view $\mathbf{F}$ as restricted to the hyperplane $\mathbf{H}$. Then for X in $\mathbf{H}$, the condition $\mathbf{F}(X) = 0$ implies $X = O$. Finally, for X in $\mathbf{H}$, we have $\mathbf{F}(-X) = \mathbf{F}(X)$. Hence $\mathbf{F}$ is a norm function on $\mathbf{H}$, and in particular, induces a norm function on any subspace of $\mathbf{H}$.

The image of Γ under L is a discrete subgroup of $\mathbf{R}^{s+1}$. Let W be the vector space generated by this image. Then $L(\Gamma)$ is a lattice (having maximal rank) in the vector space W, of dimension r. Let $B_1, \ldots, B_r$ be a basis of $L(\Gamma)$ over $\mathbf{Z}$. For any vector X in W we can write

$$X = y_1 B_1 + \cdots + y_r B_r.$$

We define

$$G(X) = \sup |y_i|.$$

Then G is a norm on W. Since two norms on a finite dimensional vector space are always of the same order of magnitude, we conclude:

Lemma 1. *We have* $\mathbf{F} \gg\ll \mathbf{G}$ *on the space* W.

Let $\alpha_1, \ldots, \alpha_r$ be free generators of Γ modulo its torsion group. For each α in Γ we can write

$$\alpha = \alpha_1^{m_1} \ldots \alpha_r^{m_r} \zeta,$$

where ζ is a root of unity. We define

$$m = m(\alpha) = \sup |m_i|.$$

Observe that our definition of $\mathbf{F}$ is such that

$$\mathbf{F}(L(\alpha)) = h_K(\alpha) = \sum_v \sup(0, v(\alpha)) .$$

From Lemma 1, we obtain:

Lemma 2. *As functions on Γ, we have $m \gg\ll h$.*

Theorem 1.1. *The equation $a\alpha + b\alpha' = 1$ has only a finite number of solutions in Γ.*

Proof. We shall reduce the proof to a statement of diophantine approximation. For concreteness, we first suppose that the set S consists only of the archimedean absolute values, so that Γ is a group of ordinary units, and we could take it to be the group of all units. The set S is the set of all archimedean absolute values. For solutions α, α' of the above equations, we then have

$$H_K(\alpha) = \prod_{v \in S} \max(1, |\alpha|_v^{n_v})$$

and similarly for α'. If there are infinitely many solutions, then passing to a subset we may assume without loss of generality that for one of the absolute values v_0 we have

$$\max_v |\alpha|_v = |\alpha|_{v_0} .$$

We then denote $|\alpha|_{v0}$ simply by $|\alpha|$. This corresponds to a definite embedding of K into the complex numbers. We then have

$$|\alpha|^N \geqslant H_K(\alpha), \text{ so } |\alpha| \geqslant H_K(\alpha)^{1/N} ,$$

where $N = [K:\mathbf{Q}]$. Note that for any infinite set of solutions α as above, the absolute value $|\alpha|$ becomes arbitrarily large.

We apply the mean value theorem to the equation $a\alpha + b\alpha' = 1$, with the log function (principal value). Since $|a\alpha|$ is large, this yields an inequality of the form

$$(*) \qquad |\log a\alpha - \log b\alpha'| \ll \frac{1}{|a\alpha|} \ll \frac{1}{|\alpha|} \ll \frac{1}{H(\alpha)} .$$

Next, we write α, α' in terms of a set of generators for Γ,

$$\alpha = \alpha_1^{m_1} \dots \alpha_r^{m_r} \zeta \quad \text{and} \quad \alpha' = \alpha_1^{m'_1} \dots \alpha_r^{m'_r} \zeta' .$$

Passing to a subset of solutions if necessary, we may assume without loss of generality that the roots of unity (ζ, ζ') are the same for all pairs of solutions (α, α') to our equation. Note that $m(\alpha) \gg\ll m(\alpha')$, since trivially $H(\alpha) \gg\ll H(\alpha')$. We let

$$u_j = \log \alpha_j .$$

Also put $u_0 = 2\pi i$. Then (*) can be written in the form

$$|q_0 u_0 + q_1 u_1 + \cdots + q_r u_r - u_{r+1}| \ll \frac{1}{H(\alpha)},$$

where $u_{r+1} = \log(a\zeta/b\zeta')$, and the coefficients q_j are integers, satisfying

$$q = \max |q_j| \ll m(\alpha) .$$

In the light of Lemma 2, we therefore obtain the inequality

$$|q_0 u_0 + q_1 u_1 + \cdots + q_r u_r - u_{r+1}| \ll \frac{1}{C^q}$$

with some constant C. Therefore any lower bound for linear combinations of algebraic numbers which contradicts this inequality will suffice to prove Theorem 1.1. The proof of such lower bounds will be given in Chapter IX, as desired.

Note that the proof effectively gives a bound for the heights of the solutions of the equation, if an effective lower bound can be given for the final diophantine inequality. The bounds in all other steps are trivially obtainable.

We now say a few words concerning the case of generalized units, i.e. when the set S contains non-archimedean absolute values. In that case, the special absolute value v_0 which we selected may be $\mathfrak{p}$-adic, corresponding to an embedding of K in the algebraic closure of $\mathbf{Q}_p$. Again we denote this absolute value without subscript. The p-adic logarithm is defined only in a neighborhood of 1 by the usual power series. Hence we need a technical device to apply Gelfond's idea. We make no attempt to be efficient, and carry out the simplest conceptual idea for this. We write $c = -b/a$ and

$$a\alpha + b\alpha' = a\alpha(1 - c\alpha'/\alpha) = 1 .$$

Then

$$|1 - c\alpha'/\alpha| = \frac{1}{|a\alpha|} \ll \frac{1}{|\alpha|} \cdot$$

This is small, and hence $c\alpha'/\alpha$ is near 1. Consequently

$$|\log c\alpha'/\alpha| \ll \frac{1}{|\alpha|} \ll \frac{1}{H(\alpha)^{1/s}}$$

if $s = \text{card } S$, and we assume that $|\alpha| = \max_{v \in S} |\alpha|_v$.

We select generators for a subgroup of finite index in Γ as follows. Let $\mathfrak{p}_1, \ldots, \mathfrak{p}_t$ be the distinct prime ideals corresponding to non-archimedean absolute values in S. A sufficiently high power of these ideals is principal, say

$$(\pi_1) = \mathfrak{p}_1^{j_1}, \ldots, (\pi_t) = \mathfrak{p}_t^{j_t}.$$

Then $\pi_1, \ldots, \pi_t$ and the ordinary unit group generate a group of finite index in Γ. Say $\mathfrak{p} = \mathfrak{p}_t$ is the prime corresponding to the selected absolute value $\| \quad \|_{v_0}$. Let

$$\Gamma_0 = \{\pi_1, \ldots, \pi_{t-1}, \text{ordinary units}\}$$

be the group generated by $\pi_1, \ldots, \pi_{t-1}$ and the ordinary units. Then some fixed power of each generator is a $\mathfrak{p}$-adic unit close to 1. Thus we may find a set of generators

$$\{\alpha_1, \ldots, \alpha_r, \pi_t\}$$

for a subgroup Γ_1 of finite index in Γ such that $\alpha_1, \ldots, \alpha_r$ are $\mathfrak{p}$-adic units close to 1.

We may then proceed as before. We put

$$u_j = \log \alpha_j,$$

where $j = 1, \cdots, r$. Then raising the power $m = (\Gamma : \Gamma_1)$, we get

$$(c\alpha'/\alpha)^m = \alpha_1^{q_1} \ldots \alpha_r^{q_r}$$

for some integers $q_1, \ldots, q_r$. Note that π_t cannot occur in the power product on the right since $c\alpha'/\alpha$ is a $\mathfrak{p}$-adic unit.

Furthermore, the equation

$$a\alpha + b\alpha' = 1$$

and the definition of the height shows that

$$H(\alpha) \gg\ll H(\alpha')$$

and we also have

$$H(c\alpha'/\alpha) \ll H(\alpha')H(\alpha) \ll H(\alpha)^2.$$

In view of Lemma 2 we find

$$\begin{aligned} |q_1 u_1 + \cdots + q_r u_r| &= m \log (c\alpha'/\alpha) \\ &\ll \frac{1}{H(\alpha)^{1/s}} \leqq \frac{1}{C^q} \end{aligned}$$

with some constant C, and where as before

$$q = \max |q_j| .$$

In this manner the diophantine statement is reduced to a $\mathfrak{p}$-adic inequality analogous to that of Baker. Appropriate lower bounds giving a contradiction for q large have been proved by Coates [Co 1]. The proof essentially follows the arguments which will be given in Chapter VIII, and one obtains a similar $\mathfrak{p}$-adic inequality corresponding to the Baker–Feldman theorem. More recently, improvements have been given by Van der Poorten [VdP 2].

We shall now make some comments concerning the effectiveness with which one can estimate the various steps in the above argument, and the uniformity which one gets over, say, number fields of bounded degree.

We let again K be our number field and let D be the absolute value of the discriminant. The image under L of the units is a lattice. Let

$$L(\alpha_1), \ldots, L(\alpha_r)$$

be the successive minimal points of this lattice, with successive minima

$$|L(\alpha_i)| = \lambda_i .$$

(For the theory of successive minima, see Davenport [Da].) We recall that these points are chosen inductively to have minimal length, subject to being linearly independent. Let R be the regulator. Siegel [Si 2] has shown that

$$\prod_{i=1}^{r} h_K(\alpha_i) \ll R ,$$

or in terms of successive minima,

$$\lambda_1 \ldots \lambda_r \leqslant R . \tag{1}$$

Furthermore,

$$R \ll (\log D)^{N-1} \sqrt{D} \tag{2}$$

where $N = [K : \mathbf{Q}]$.

The successive minimum points do not form a basis of the lattice, but they almost do, and it is also known from the theory of successive minima that there is a basis such that in the orthonormal system associated with the sequence of minimum points, the i-th element of the basis has euclidean norm $\gg\ll \lambda_i$, and also has i-th component $\gg\ll \lambda_i$. Thus one can either work with such a basis of units, or work with the subgroup Γ of the units generated by the Siegel units α_i. Let us do the latter, to minimize knowledge about minimum points.

We consider the equation

$$a\alpha + b\alpha' = 1$$

where $\alpha, \alpha' \in \Gamma$. There is only a finite number of elements of bounded height in K, and a bound can easily be given, effectively depending only on the degree (bounding the coefficients of an irreducible equation over the integers). In practice, one first gives a bound for the coefficients a, b at finite primes. They can then be multiplied with elements of Γ to have absolute value at $v_1, \ldots, v_r$ bounded in terms of the discriminant. Then $H_K(\alpha)$ can be assumed to be large with respect to $|a|_{v_i}$ and $|b|_{v_i}$ for $i = 1, \ldots, r$. Thus as before $|a\alpha|$ can be viewed as large, and we have the inequality

$$|\log a\alpha - \log b\alpha'| \ll \frac{1}{|a\alpha|}$$

$$\ll \frac{H_K(a)}{H_K(\alpha)^{1/N}}$$

if we select the absolute value as before to be maximal. We don't want to view a, b as fixed any more, so we have to keep track of them. We proceed as before, but are more careful with the coefficients q_j. The expression

$$b\alpha'/a\alpha$$

is close to 1. We have

$$\log b\alpha'/a\alpha = \log b/a + \log \alpha'/\alpha + \text{multiple of } 2\pi i\,.$$

We pick all log values in the fundamental domain. The multiple of $2\pi i$ is then $\ll |\log |b/a|| \ll A$, where we put

$$A = \max\,(H_K(a), H_K(b))\,.$$

Hence we get the inequality

$$|q_0 u_0 + q_1 u_1 + \cdots + q_r u_r - u_{r+1}| \ll \frac{A}{C^{\sum m_i \lambda_i}}$$

with a constant C depending only on N, and:

$$|q_0| \ll \log A$$

$$|q_i| \ll m(\alpha) + m(\alpha'), \qquad \text{for } i = 1, \ldots, r\,.$$

In practice, $\max |m_i| \lambda_i$ is comparatively large with respect to A, so that the denominator of the right hand side dominates the numerator.

To get effective estimates, one needs a strong inequality on the left. In the next part we prove such inequalities. If we use that of Baker [Ba 3], then the left hand side is seen to satisfy

$$\frac{1}{C_1^{(\log q + \log A)(\lambda_1 \cdots \lambda_r)^{1+\varepsilon}}} \leqslant |q_0 u_0 + \cdots + q_r u_r - u_{r+1}| \, .$$

Comparing with the right inequality yields effective bounds. However, the presence of the product seems here unnatural, and it would be far preferable to have an inequality as discussed in the Lang–Waldschmidt conjecture, see the introduction to Chapters X and XI. This would allow for a much finer comparison with the right hand side in the applications, as discussed in the next section.

In the preceding discussion we limited ourselves to ordinary units for concreteness. Waldschmidt has pointed out to me that by using the proof of Theorem 1 in [VdP–L 2], one can get an inequality comparing $m(\alpha)$ and $h(\alpha)$ in general as follows. Let A_j be the maximum of the absolute values of the coefficients in a minimal polynomial for α_j. The inequality $h \gg m$ in Lemma 2 is trivial. On the other side, Waldschmidt writes me that one has:

$$m(\alpha) \leqslant D^{(r+1)/2}(3rD)^r(21D \log (6D))^{\min(r+1,D)} \left(\prod_{j=1}^{r} \log A_j \right) h(\alpha) \, .$$

The rest of this chapter consists in exhibiting various cases of integral diophantine problems which can be reduced effectively to the equation in units of § 1. Proofs of the reduction are given in each case, and in particular, the following equations are shown to have only finitely many solutions in the integers of a number field:

Elliptic, hyperelliptic, and superelliptic equations of the form

$$y^n = f(x)$$

where f is a polynomial, and the equation has genus $\geqslant 1$.

Equations of Thue–Siegel type,

$$f(x, y) = \lambda$$

where f is a homogeneous polynomial of degree $\geqslant 3$.

Certain equations defining curves of genus 0.

For each of these cases, the arguments of the following sections prove:

Theorem 1.2. *Given the affine curve V defined by one of the above equations over a number field K. There exists a finite extension K' of K, which can be determined explicitly depending on V, and a finite number of equations*

$$E_i \colon a_i T + b_i T' = 1, \quad i = 1, \ldots, m$$

with coefficients a_i, b_i in K', such that the solutions of V in $\mathfrak{o}_K$ correspond to a subfamily of solutions of the equations E_i in units in K'. The correspondence can be exhibited effectively.

§ 2. Reduction of Integral Points to the Unit Equation by Siegel's Method

We consider the elliptic curve defined by the ordinary equation

$$y^2 = x^3 + ax + b .$$

At first, a, b lie in a number field K. We are interested in the integral points on this curve. More generally, a set J of points (x, y) in A_K is said to be **quasi-integral** if there exists a positive integer d such that $dx, dy \in \mathfrak{o}_K$ for all $(x, y) \in J$. By making a simple transformation on the curve, a set of quasi integral points can always be transformed into a set of integral points, and the coefficients a, b may be assumed integral. Indeed, let us write

$$a = a'/d \quad \text{and} \quad b = b'/d ,$$

where d is taken to be an integer which is both a square, a cube, and sufficiently divisible to be a denominator for a, b and the points in J. Furthermore, a', b' are integers such that a' is also divisible by a denominator for points in J. Write

$$d = d_0^2 , \qquad d = d_1^3 , \quad \text{and} \quad a' = d_1 a'' .$$

Solutions of the given equation are equivalent with solutions of the transformed equation

$$(d_0 y)^2 = (d_1 x)^3 + a''(d_1 x) + b' .$$

The finiteness of integral points on the latter implies the finiteness of quasi-integral points on the former equation. *From now on, we assume $a, b \in \mathfrak{o}_K$.*

We let J be the set of points $P \in A_K$ such that $x(P) \in \mathfrak{o}_K$. This is the set of integral points. We want to show how the integral points give rise to solutions of the unit equation discussed in the preceding section. We write our equation in the form

$$y^2 = h(x) = (x - e_1)(x - e_2)(x - e_3) .$$

Making a finite extension if necessary, we may assume $e_1, e_2, e_3 \in K$. For each integral solution (x, y) with $x \neq e_i$, the algebraic integers $x - e_i$ are quasi-relatively prime: the only common factors are those dividing the differences $e_i - e_j$ $(i \neq j)$.

Thus we have ideal factorizations

$$(x - e_i) = \mathfrak{a}_i^2 \mathfrak{b}_i$$

where $\mathfrak{b}_i = \mathfrak{b}_{i,x}$ can range over only a finite number of ideals, and $\mathfrak{a}_i = \mathfrak{a}_{i,x}$ depends on x. Let $\mathfrak{c}$, $\mathfrak{c}'$ be selected from a fixed set of ideal representatives for the ideal classes, and such that $\mathfrak{c}\mathfrak{a}_i$ and $\mathfrak{c}'\mathfrak{b}_i$ are principal. Multiplying the above factorization by $\mathfrak{c}^2\mathfrak{c}'$ yields equations

$$x - e_i = z_{i,x}^2 w_{i,x}$$

where the algebraic numbers $w_{i,x}$ can range only over a finite number of values. We consider the set of those points (x, y) giving rise to the same system of values (w_1, w_2, w_3), and write

$$w_i = t_i^2$$

so that t_i is at most quadratic over K. Then

$$x - e_i = z_{i,x}^2 t_i^2 .$$

Subtracting shows that for $i \neq j$ the triple $(z_{1,x}, z_{2,x}, z_{3,x})$ satisfies the equation

$$e_i - e_j = (z_j t_j + z_i t_i)(z_j t_j - z_i t_i) . \tag{1}$$

This shows that the factors on the right divide the fixed values $e_i - e_j$ on the left. Therefore there exist a finite number of elements $e_{i,j,x}$ in $K(t_1, t_2, t_3)$ such that

$$z_{j,x} t_j - z_{i,x} t_i = e_{i,j,x} u_{i,j,x}$$

where $u_{i,j,x}$ are units. Again restricting ourselves to those x having the same system of values $e_{i,j}$ we see that

$$(z_{1,x}, z_{2,x}, z_{3,x})$$

satisfy the relations

$$\begin{aligned} z_1 t_1 - z_2 t_2 &= \beta_1 u_1 \\ z_2 t_2 - z_3 t_3 &= \beta_2 u_2 \\ z_3 t_3 - z_1 t_1 &= \beta_3 u_3 \end{aligned} \tag{2}$$

where $(\beta_1, \beta_2, \beta_3)$ is a fixed set of algebraic numbers, and u_1, u_2, u_3 are units. The sum of these three expressions is equal to 0, namely

$$\beta_1 u_1 + \beta_2 u_2 + \beta_3 u_3 = 0 .$$

Dividing by $\beta_3 u_3$, say, yields solutions in units of the linear equation considered in the preceding section, and therefore has only a finite number of solutions. Thus the point (u_1, u_2, u_3) can have only a finite number of values up to a proportionality factor, which is itself a unit.

The system of equations (1) *and* (2) *can have only one solution when the numbers* t_i, β_i, u_i $(i = 1, 2, 3)$ *are fixed.*

Indeed, the linear equations have rank 2, and from (1), we see that the values

$$z_j t_j + z_i t_i$$

are also fixed, so this third linear relation determines the triple (z_1, z_2, z_3) completely.

Changing the u_i by a common factor u then changes the solution by this same factor, and from (1) we see that $u^2 = 1$, so $u = \pm 1$.

Hence finally we conclude that the equations

$$x - e_i = z_i^2 t_i^2$$

have only a finite number of solutions, as desired.

Remark. The argument works just as well for the hyperelliptic equation

$$y^2 = h(x)$$

where h has at least three distinct roots, using only those roots to carry out the above steps.

Both for its own sake, as in the study of equations

$$y^2 = x^3 + k\,,$$

and for other applications, it is desirable to have uniformity properties concerning the size of solutions in terms of the coefficients. Baker was the first to obtain such results [Ba 4], improved by Stark [St 2] using Siegel's estimates for units. The Siegel method of this section allows one to work directly with the equation, rather than passing through the auxiliary Thue–Siegel equation (discussed also below).

An interesting application would be for the comparison of the discriminant

$$\Delta = -16(4a^3 + 27b^2)$$

when a, b range over the integral coefficients of a Weierstrass equation of a minimal model over $\mathbf{Z}$. This amounts to the above equation, for which M. Hall has conjectured that

$$|y^2 - x^3| \gg \max\{|x|^{\frac{1}{2}}, |y|^{\frac{1}{3}}\}\,.$$

See his article in Computers in Number Theory, Oxford conference, Academic Press, 1971. Otherwise not much is known beyond the weak lower bound given by Stark [St 2]. A diophantine inequality of the type discussed at the end of the preceding section might allow an improvement like

$$|\Delta| \gg \max\{|a|, |b|\}^{\kappa}$$

with some positive κ. See the introduction to Chapters X and XI.

In this connection, one could give a general setting for some of Demjanenko's ideas [De 5], [De 6], and I would conjecture that on a quasi-minimal model of an elliptic curve over a number field K, the number of integral points is bounded by a number depending only on the rank of the Mordell–Weil group (and of course K). However, to pursue these ideas would require knowing a lower bound for the Néron–Tate height of points of infinite order, and such bounds are not known. Furthermore, even in the special case considered by Demjanenko in [De 6], I cannot understand his Lemma 4, which seems to overlook completely what are the essential difficulties in dealing with this type of problem.

§ 3. Chabauty's Method

Chabauty [Ch] gave another method of reducing the analysis of integral points on elliptic curves to the unit equation (although he did not phrase it in terms of "units", cf. [KL]). *In this section* we assume that the reader is acquainted with the basic theory of function fields in one variable (algebraic curves), and integral closure.

Consider the elliptic curve A defined by the usual equation

$$y^2 = h(x) = (x - e_1)(x - e_2)(x - e_3).$$

We take (x, y) as a generic point, and write

$$(x, y) = 2(X, Y).$$

We suppose A defined over the number field K, taken to be sufficiently large to contain all e_i. Let

$$u = x - e_1 \quad \text{and} \quad u' = x - e_2.$$

Then u, u' are functions on the curve, and

$$u - u' = e_2 - e_1.$$

Let P_1, P_2 be the points of order 2 on A corresponding to putting $x = e_1$ and $x = e_2$. Then the divisors of the two functions are given by

$$(u) = 2(P_1) - 2(O) \quad \text{and} \quad (u') = 2(P_2) - 2(O).$$

Let

$$f: A \to A$$

be the covering obtained by "multiplication by 2". Then f is unramified. Let

$$R = K[x, y] \quad \text{and} \quad F = K(x, y)$$

be the affine ring of functions on the base curve A, and the function field respectively. Let R_2 be the integral closure of R in the function field

$$F_2 = K(X, Y)$$

of the covering curve. Then the functions u, u' are units in R_2 because their only zeros and poles are at points lying above the origin O on the base curve under the covering map f. We may write the ring R_2 as

$$R_2 = K[Z_1, \ldots, Z_n]$$

for some elements Z_j which are integral over R. Each Z_j satisfies an equation of the form

$$cZ^d + \varphi_{d-1}Z^{d-1} + \cdots + \varphi_0 = 0\,,$$

where $c \in K$, and $\varphi_i \in K[x, y]$. Clearing denominators multiplying with an appropriate ordinary positive integer, we may assume that $c \in \mathbf{Z}$, $c > 0$, and $\varphi_i \in \mathfrak{o}_K[x, y]$. This shows that if P is an integral point, i.e. $x(P) \in \mathfrak{o}_K$, $y(P) \in \mathfrak{o}_K$, then the points in its inverse image $f^{-1}(P)$ are quasi-integral. Since the functions $u \circ f$ and $u' \circ f$ are units, their values at points in the inverse image $f^{-1}(P)$ are **quasi-units**, i.e. differ from units by multiplication with elements having bounded ideal factorization.

To conclude the proof, it is now necessary to know that the points in the inverse images $f^{-1}(P)$ must be rational over a *finite* extension of K. This is achieved by a theorem of Chevalley Weil [Ch-W].

Theorem 2.1. *Let f: $W \to V$ be an unramified covering of a complete non-singular curve by another, defined over a number field K. There exists a positive integer c having the following property. For any point P of V in the algebraic closure of K, the relative discriminant of $K(f^{-1}(P))$ over $K(P)$ divides c.*

The original proof proceeds by writing down the fact that the ideal generated in affine rings for V by the discriminants of generators for the integral closure are unit ideals, thus giving a generic upper bound for the discriminant, which implies bounds for the discriminant in any specialization at a point. For the present application, it is not necessary to have the full strength of the theorem, we need only the fact that the extensions

$$K(f^{-1}(P))$$

are unramified outside a finite set of primes, for then we can use the elementary fact that there is only a finite number of extensions of K of bounded degree, unramified outside a given finite set of primes. But the existence of such a finite set of primes is obvious if one is willing to assume the formalism of reduction mod $\mathfrak{p}$. For all but a finite number of primes $\mathfrak{p}$ in K, the covering reduces to a covering

$$f_{\mathfrak{p}}\colon W_{\mathfrak{p}} \to V_{\mathfrak{p}}$$

which is also unramified, whence it follows that for such $\mathfrak{p}$, the extension $K(f^{-1}(P))$ is unramified over $\mathfrak{p}$, as desired.

We have picked the Weierstrass model to start with for convenience. In fact, Chabauty himself argues more abstractly, selecting his functions u, u' characterized by the divisorial conditions

$$(u) = 2(P_1) - 2(O) \quad \text{and} \quad (u') = 2(P_2) - 2(O)$$

on any model of the curve. By selecting the origin O at infinity on a given affine model, he obtains the bound for integral points on an arbitrary model. For clarity, we extract this reduction separately in the next section, and make further historical comments in that connection.

§ 4. Reduction to the Weierstrass Form

Let A be an elliptic curve defined over a number field K, and let $K(A)$ be its function field. Let φ be a non-constant function in $K(A)$. One may ask about the integral points with respect to φ, i.e. those points $P \in A_K$ such that $\varphi(P) \in \mathfrak{o}_K$ (or more generally, $\varphi(P) \in R$ where R is a finitely generated subring of K). There is a standard birational reduction of any curve of genus 1 to Weierstrass form, known explicitly in any abstract context since F. K. Schmidt proved the Riemann–Roch in general. For an early reference, cf. Hasse's comments on Mahler's paper [Mah 4]. If this reduction is made by using as origin one of the poles of φ, then it transforms φ-integral points to quasi-integral points on the Weierstrass model, and thus achieves the desired reduction. We repeat the argument briefly.

Let O be a pole of φ, which we may assume rational over K after making a finite extension if necessary. Let x, y be the Weierstrass functions such that x has a pole of order 2 at O and y a pole of order 3, satisfying an equation

$$y^2 = x^3 + ax + b$$

with $a, b \in K$. Any point of A which is not a pole of φ is also not a pole of x. Consequently x satisfies an integral equation

$$x^d + g_{d-1}(\varphi)x^{d-1} + \cdots + g_0(\varphi) = 0\,,$$

where $g_i(\varphi) \in K[\varphi]$ are polynomials in φ. Multiplying by some positive integer, we may assume without loss of generality that the equation has the form

$$cx^d + g_{d-1}x^{d-1} + \cdots + g_0 = 0\,,$$

where $c \in \mathbf{Z}$, $c \neq 0$, and $g_i \in \mathfrak{o}_K[\varphi]$. This shows that any φ-integral point induces a point which is quasi-integral on the Weierstrass model with coordinates (x, y), and reduces the proof of finiteness to the Weierstrass form.

Historically, the type of language used since Weil's thesis and Siegel's results on integral points [Si 1], involving primes in function fields in one variable, as for instance also in Hasse's paper on Siegel's theorem for curves of genus 1 [Ha], allowed a direct passage from integrality of functions with respect to primes of the function field, to quasi-integrality of values in the algebraic number field. Cf Weil's theory of "distributions" as for instance in [We 1], [Si 1], or [L 1], Chapter VII, § 3, reformulated and extended by Néron as the theory of "quasi-functions", [Ne 2]. We recall here the statement (but not the proof).

Let V be a complete non-singular curve defined over a number field K. To each prime rational divisor T of V over K, and to each point $P \in V_K$, one can associate an integral ideal $\mathfrak{a}_T(P)$ having the following property. If φ is a function in the function field $K(V)$, and its divisor is

$$(\varphi) = \sum m_i T_i$$

where the T_i are distinct prime divisors of V rational over K, then for any point $P \in V_K$ which is not a zero or pole of φ we have

$$\prod \mathfrak{a}_{T_i}(P)^{m_i} \prec (\varphi(P)) \prec \prod \mathfrak{a}_{T_i}(P)^{m_i}\,.$$

If T and T' are distinct prime divisors, then

$$\text{g.c.d.}\,[\mathfrak{a}_T(P), \mathfrak{a}_{T'}(P)] \prec (1)\,.$$

We have used the notation $\mathfrak{a} \prec \mathfrak{a}'$ to mean the following. There exists an integral ideal $\mathfrak{b}$ depending only on φ such that

$$\mathfrak{a} \text{ divides } \mathfrak{a}'\mathfrak{b}\,.$$

The first assertion may be stated by saying that the functions

$$P \mapsto (\varphi(P)) \quad \text{and} \quad P \mapsto \prod \mathfrak{a}_{T_i}(P)^{m_i}$$

are quasi-equal. The second assertion may be stated by saying that $\mathfrak{a}_T$ and $\mathfrak{a}_{T'}$ are quasi-relatively prime, i.e. relatively prime up to bounded factors. We have stated

Weil's theorem for points in K, but as usual, it is valid uniformly for algebraic points. For genus 0, a proof can easily be given, see § 6.

Baker and Coates [BC], [Co 2] have given explicit determinations of the reduction of integrality to quasi-integrality, and explicit ways of finding functions with given zeros and poles on a curve (i.e. explicit Riemann–Roch theorem). Before Baker proved his inequality on logarithms of algebraic numbers, no attention had been paid to being sure that all steps in the proof of finiteness were constructive, because the only known arguments were Siegel's and Chabauty's. Siegel's arguments using Thue–Siegel were not constructive. His special argument for hyperelliptic curves, even though passing through the unit equation, immediately climbed back to a Fermat equation, and used Thue–Siegel again. As for Chabauty, his paper was unfortunately somewhat disregarded, partly because in the latter part he gave an argument claiming to extend his method to all curves. At the time this was not understood (the reviews in *Math Reviews* are non-committal), and it turned out to be false. The matter is discussed in [KL]. After Baker proved his inequality it became again important to go through the old proofs and examine them in the new light. It then turned out that the reduction of the diophantine problem to the unit equation

$$a\alpha + b\alpha' = 1$$

by Siegel's or Chabauty's methods were quite constructive. The new idea of Gelfond to use logarithms of algebraic numbers and a diophantine inequality for linear combinations of these could then be carried out because of Baker's method to handle such combinations.

§ 5. The Thue–Siegel Curve

We describe an argument of Siegel showing how to handle an equation

$$\prod_{i=1}^{n} (X - \alpha_i Y) = \lambda$$

where α_i, λ are elements of a number field K. The α_i need not be distinct, but we assume that there are at least three distinct such numbers, say $\alpha_1, \alpha_2, \alpha_3$. In [Si 1], Part 2, § 1 Siegel shows how to reduce solving this equation in integers to a finite number of equations

$$au + bu' = 1$$

in units u, u' as follows.

To begin with, without loss of generality, we may assume that α_i, λ are integral. Indeed, let c be a positive integer which is a denominator for all α_i and λ. Multiplying the above equation by c^n reduces it to

$$\prod (cX - c\alpha_i Y) = \lambda c^n .$$

Let $X' = cX$, $\alpha_i' = c\alpha_i$, and $\lambda' = \lambda c^n$. If we can bound the solutions of the equation

$$\prod (X' - \alpha_i' Y) = \lambda'$$

in $\mathfrak{o}_K$, then we also get a bound for the solutions of the original equation in $\mathfrak{o}_K$.

Without loss of generality, we therefore assume that α_i, λ are integral.

For any integral solution X, Y the factors $X - \alpha_i Y$ then divide λ, and consequently lie in a finite number of cosets of the units. We now use the Siegel identity

$$\frac{\alpha_3 - \alpha_1}{\alpha_2 - \alpha_1} \frac{X - \alpha_2 Y}{X - \alpha_3 Y} + \frac{\alpha_2 - \alpha_3}{\alpha_2 - \alpha_1} \frac{X - \alpha_1 Y}{X - \alpha_3 Y} = 1 .$$

Each ratio

$$\frac{X - \alpha_i Y}{X - \alpha_j Y}$$

lies in only a finite number of cosets of the units, and therefore the original equation gives rise to a finite number of equations of type

$$au + bu' = 1$$

in units. If we know that these have only a finite number of solutions, then we conclude that the ratios above are finite in number. For a fixed pair of elements (γ, γ') in K, the equations

$$\frac{X - \alpha_2 Y}{X - \alpha_3 Y} = \gamma \quad \text{and} \quad \frac{X - \alpha_1 Y}{X - \alpha_3 Y} = \gamma'$$

then determine the pair (X, Y) up to a proportionality factor. Thus all solutions of the original equation corresponding to a fixed pair (u, u') are of type

$$(X, Y) = \rho(X_0, Y_0) , \quad \rho \in K ,$$

where X_0, Y_0 are fixed. Thus the original equation yields

$$\rho^n \prod_{i=1}^{n} (X_0 - \alpha_i Y_0) = \lambda .$$

This shows that ρ can have only a finite number of values and concludes the proof of the reduction of the Thue–Siegel equation to the unit equation treated in § 1.

§ 6. Curves of Genus 0

Siegel applied the analysis of the preceding section to curves of genus 0 as follows.

Theorem 6.1. *Let $\varphi(t)$ be a rational function with coefficients in a number field K. If φ has at least three distinct poles, then there are only finitely many $t \in K$ such that $\varphi(t) \in \mathfrak{o}_K$ (the ring of integers in K).*

Proof. After a fractional linear transformation, we may assume that the zeros and poles of φ do not lie at 0 or infinity, and thus φ admits a factorization

$$\varphi(t) = c\frac{f(t)}{g(t)}$$

where c is a leading coefficient, and

$$f(t) = \prod_{i=1}^{n} (t - \beta_i), \qquad g(t) = \prod_{i=1}^{n} (t - \alpha_i)$$

are polynomials of the same degree and relatively prime. Note that one can also write

$$\varphi(t) = c\frac{\prod (1 - \beta_i/t)}{\prod (1 - \alpha_i/t)}$$

since f, g have the same degree. Making a finite extension of K if necessary, we may assume that the roots α_i, β_i lie in K.

Any non-zero element t in K can be written as a quotient of integral elements,

$$t = X/Y,$$

such that X, Y are quasi-relatively prime (their g.c.d. divides a fixed number). Let us write

$$g(X, Y) = \prod (X - \alpha_i Y)$$
$$f(X, Y) = \prod (X - \beta_i Y).$$

Since f, g are relatively prime, there exist relations

$$A_1(X, Y)f(X, Y) + B_1(X, Y)g(X, Y) = Y^n$$
$$A_2(X, Y)f(X, Y) + B_2(X, Y)g(X, Y) = X^n$$

where A_1, A_2, B_1, B_2 are (homogeneous) polynomials. If X, Y are obtained from a solution t such that $\varphi(t)$ is integral, then dividing by $g(X, Y)$ shows that $g(X, Y)$

divides Y^n/d and X^n/d where d is a positive integer denominator for the coefficients of A_1, A_2, B_1, B_2. Since X, Y are quasi relatively prime, it follows that

$$g(X, Y) = \lambda u ,$$

where λ ranges over a finite number of elements of K, and u ranges over units.

Replacing K by the finite extension obtained by adjoining n-th roots of units, we may write $u = w^n$ for some unit w. Since $g(X, Y)$ is homogeneous of degree n, the above equation to be solved in integral X, Y is equivalent to the equation

$$g(X, Y) = \lambda$$

to be solved in integral X, Y. This reduces the case of genus 0 to the case treated in the preceding section.

§ 7. Applications to Curves of Higher Genus

In this section we follow Kubert–Lang [KL] to show how the problem of solving certain equations of higher genus in integers can be reduced to elliptic curves.

Consider the hyperelliptic curve V defined by

$$y^2 = f(x) = (x - \alpha_1)\ldots(x - \alpha_n) ,$$

where $n \geqslant 4$ and the roots are distinct. Then the elliptic curve A defined by

$$y^2 = (x - \alpha_1)(x - \alpha_2)(x - \alpha_3)(x - \alpha_4)$$

is ramified only at α_i for $i = 1, \ldots, 4$. Let W be the pull back of A over V, written $W = A \otimes V$. By definition, W is the curve in the product space consisting of pairs (P, Q) with $P \in V$ and $Q \in A$ projecting on the same point on the projective x-line. Then W is unramified over V, and by the Chevalley–Weil theorem already used in Chabauty's theorem, it follows that integral points of V in K give rise to quasi-integral points of W in a fixed finite extension of K, whence to quasi-integral points on A in such an extension. This reduces the bounding of integral points on V to quasi-integral points on A. The diagram is as follows.

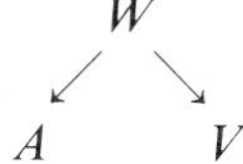

More generally, consider the superelliptic curve V defined by the equation

$$y^n = f(x) = (x - \alpha_1)\ldots(x - \alpha_r) .$$

For instance, the Catalan equation $y^n - x^r = 1$ is of this type (see Chapter IX). We assume for simplicity that the numbers α_i are all distinct. Suppose also for simplicity that n, r are distinct prime numbers, and $n \geqq 3$. Then the curve V' defined by

$$y^n = (x - \alpha_1)(x - \alpha_2)$$

is ramified only at α_1, α_2 and infinity, of the same order as V. Hence the pull back $V' \otimes V$ is unramified over V. Furthermore, V' is hyperelliptic since x is quadratic over the function field in y (the role of x, y is reversed here from what it was in the preceding discussion!). Therefore by the hyperelliptic discussion, we can find an elliptic curve A such that $A \otimes V'$ is unramified over V'.

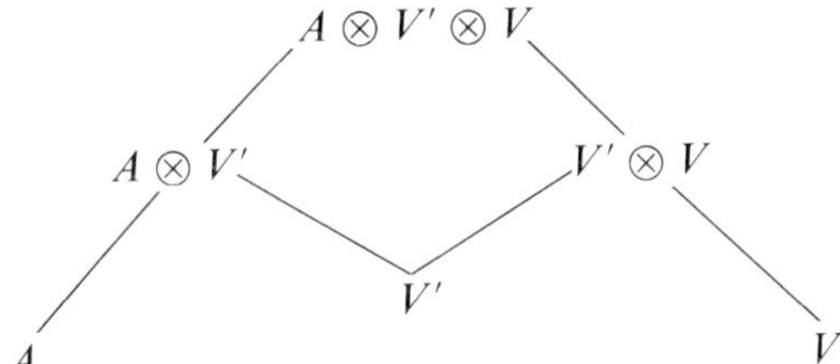

This reduces the diophantine problem from V to A.

For the reduction of other types of curves, e.g. modular curves, to the unit equation $au + bu' = 1$, we refer to [KL].

§ 8. Reduction to Inequalities on Elliptic Logarithms

By means of the Weierstrass functions, we have a complex analytic map

$$\exp : \mathbf{C} \to A_{\mathbf{C}}$$

from the complex numbers (modulo the period lattice) onto the complex points on the elliptic curve. If $P \in A_{\mathbf{C}}$ and $P = \exp u$, we also write

$$u = \log P ,$$

and call u an **elliptic logarithm**. We choose u in a fundamental parallelogram. This corresponds to selecting principal values for the ordinary logarithm.

We shall deal with the case when A is defined over a number field K, say with Weierstrass coordinate functions (x, y). We let $\varphi \in K(x, y)$ be a non-constant function on A. We wish to get a bound on the heights of points $P \in A_K$ such that

$$\varphi(P) \in \mathfrak{o}_K .$$

In Chapter IV, we took the height with respect to the x-coordinate. Using routine inequalities, it is easily shown that if we define

$$h_{\varphi}(P) = h(\varphi(P)),$$

then

$$h_{\varphi} \gg\ll h_x .$$

We leave the proof to the reader. For a φ-integral point P we have

$$H_{K,\varphi}(P) = \prod_{v \in S_\infty} \max(1, |\varphi(P)|_v^{n_v}) .$$

Consider those φ-integral points such that $\max(1, |\varphi(P)|_v)$ for $v \in S_\infty$ occurs for a fixed absolute value v_0, and write

$$| \ | = | \ |_{v_0} .$$

We take the complex embedding of K corresponding to this absolute value. Then we get

$$|\varphi(P)| \geqslant H_{\varphi}(P)$$

where

$$H_{\varphi}(P) = H_{K,\varphi}(P)^{1/N} \quad \text{and} \quad N = [K : \mathbf{Q}] .$$

Let Γ be a subgroup of A_K, generated by points $P_1, \ldots, P_r$. Consider the set of φ-integral points P such that $P \in \Gamma$. Let

$$u_j = \log P_j ,$$

where the log is taken with respect to the above chosen complex embedding. Any sequence of such points P whose height tends to infinity must approach a pole of φ, which we denote by P_{r+1}, and this pole is algebraic. Let $u_{r+1} = \log P_{r+1}$. Write

$$P = q_1 P_1 + \cdots + q_r P_r$$

so that

$$\log P = q_1 u_1 + \cdots + q_r u_r \ (\text{mod periods}) .$$

There exist integers q', q'' such that the complex point

$$u = q_1 u_1 + \cdots + q_r u_r + q'\omega_1 + q''\omega_2$$

is near u_{r+1} for the above points P. If the function on $\mathbf{C}$ given by

$$\varphi(\exp(z))$$

has a pole of order m at u_{r+1}, then we have the estimate for the distance,

$$|\varphi(P)| \ll \frac{1}{|u - u_{r+1}|^m}$$

whence

$$|u - u_{r+1}|^m \ll \frac{1}{|\varphi(P)|} \leqslant \frac{1}{H_\varphi(P)}.$$

By the quadraticity of the height, and the trivial comparison between h_φ and h_x, we find an inequality

$$|q_1 u_1 + \cdots + q_r u_r + q'\omega_1 + q''\omega_2 - u_{r+1}| \leqslant C^{-q^2}$$

where C is some constant depending on φ, and

$$q = \max\{|q_j|, |q'|, |q''|\}.$$

Therefore any inequality on the left showing that linear combinations of elliptic logarithms are

$$\geqslant e^{-\tau(q)}$$

where $\tau(q) = o(q^2)$ will give the desired contradiction which implies a bound for the heights of φ-integral points.

As with ordinary logarithms, we can also carry out the same argument p-adically. Suppose that integrality is taken with respect to the ring R of elements of K which are $\mathfrak{p}$-integral for $\mathfrak{p} \notin S$, where S is a finite set of absolute values containing the archimedean ones. For φ-integral points P we then have

$$H_{K,\varphi}(P) = \prod_{v \in S} \max(1, |\varphi(P)|_v^{n_v}).$$

We consider those φ-integral points such that $\max(1, |\varphi(P)|_v)$ for $v \in S$ occurs for a fixed absolute value v, which we assume is $\mathfrak{p}$-adic, and write $|\ | = |\ |_v$. We have the same inequality

$$|\varphi(P)| \geqslant H_\varphi(P)$$

as before. We have seen that to prove the finiteness of integral points (i.e. R-integral in the present case) it suffices to do it when $\varphi = x$ is a Weierstrass function. We find it convenient to assume this now because we have already discussed the $\mathfrak{p}$-adic analytic behavior of x in Chapter III. We know that x has a pole only at the origin O. It follows that for $H(P)$ tending to infinity, the points P have to come arbitrarily close to this pole.

Let Γ be a subgroup of A_K, generated by points $Q_1, \ldots, Q_r$. There exists a positive integer m such that $mQ_j = P_j$ lies in a neighborhood of O where x, y have $\mathfrak{p}$-adically convergent power series expansion giving a local exponential parametrization of the curve in a neighborhood of the origin. We may then write

$$P_j = \exp u_j$$

where exp is the p-adic exponential, defined on a neighborhood of 0 in $\mathbf{C}_p$. If we write one of our points P in the form

$$P = q_1 Q_1 + \cdots + q_r Q_r$$

then

$$mP = q_1 P_1 + \cdots + q_r P_r = \exp(q_1 u_1 + \cdots + q_r u_r),$$

and $|x(mP)| \geqslant |x(P)|$. Consequently we get

$$|q_1 u_1 + \cdots + q_r u_r|^2 \gg\ll \frac{1}{|x(mP)|} \leqslant \frac{1}{|x(P)|} \leqslant \frac{1}{H(P)} \leqslant \frac{1}{C^{q^2}}$$

where $q = \max |q_j|$, and C is a suitable constant depending on the generators for Γ and the quadratic form for A. Thus again any inequality from the left contradicting this upper bound will conclude the proof of the finiteness theorem.

Appropriate inequalities for this purpose can be proved following the exact pattern of proof of Coates–Lang given in Chapter X. Cf. Bertrand [Be 4].

Appendix

There has been brute force working out of the explicit constants which come out of the methods of proof currently available (see part two), to get actual upper bounds for the solutions of classical diophantine equations. *At the insistence of Michel Waldschmidt*, I enclose his brief table of references to the current literature. I emphasize that these estimates do not reflect the deeper structure exhibited by these constants, depending on the geometry of numbers, Birch–Swinnerton Dyer conjecture, etc., which ultimately will give much lower and finer bounds for such solutions.

(a) *Curves of genus* 1.
$y^2 = x^3 + k$, with integral k, x, y.

$$\max\{|x|, |y|\} < \exp(C_\varepsilon |k|^{1+\varepsilon})$$

Ref: [St 2]

$y^2 = ax^3 + bx^2 + cx + d$, with integral a, b, c, d, x, y.

$$\max\{|x|, |y|\} < \exp(10^6 H)^{10^6}$$

Ref: [Ba 1], p. 45

$f(x, y) = 0$, f absolutely irreducible polynomial in $\mathbf{Z}[x, y]$ such that the curve $f = 0$ has genus 1. Integral x, y.

$$\max\{|x|, |y|\} < \exp\exp\exp\{2H^{10^{n^{10}}}\}$$

where $n = \deg f$, and H is the maximum of the absolute values of the coefficients of f (height of f).
Ref: [Ba 1], p. 45

(b) *Hyperelliptic equation*
$ay^2 = f(x)$, integral $a, f \in \mathbf{Z}[X]$, monic, at least three simple roots, $x, y \in \mathbf{Z}$.

$$\log \max\{|x|, |y|\} < C_{n,\varepsilon} |a|^{24n^2} D^{(48+\varepsilon)(n^3+3n^2)} (\log H)^{1+\varepsilon}$$

where $\varepsilon > 0$, $D =$ absolute value of discriminant of f, $H =$ height of f, $n = \deg f$.
Ref: Sprindzuck, Acta Arithm. **30** (1976) pp. 95–108

(c) *Superelliptic equation*
$y^m = f(x)$, $m \geqslant 3$, $f \in \mathbf{Z}[x]$ with at least two simple roots, integral x, y.

$$\max\{|x|, |y|\} < \exp\exp\{(5m)^{10} n^{10n^3} H^{n^2}\}$$

where $n = \deg f$ and $H =$ height of f.
Ref: Baker, Bounds for the solutions of the hyperelliptic equation, Proc. Cambridge Phil. Soc. **65** (1969) pp. 439–444

(d) *Thue equation*
$(x - \alpha_1 y)\ldots(x - \alpha_n y) = \lambda$, with $\alpha_1, \ldots, \alpha_n, \lambda \in \mathfrak{o}_K$, and $\alpha_1, \alpha_2, \alpha_3$ distinct.

$$\max\{\|x\|, \|y\|\} < \exp(dH)^{(10d)^5},$$

where $H =$ maximum of heights of the irreducible polynomials of $\alpha_1, \ldots, \alpha_n, \lambda, \xi$, and ξ is a generator of K over $\mathbf{Q}$.
Ref: [Ba 4]

$f(x, y) = m, f \in \mathbf{Z}[x, y]$ irreducible binary form of degree $n \geqslant 3$, m integer $\neq 0$, $x, y \in \mathbf{Z}$.

$$\max \{|x|, |y|\} \leqslant m^C \quad \text{with } C = C(f)\,.$$

Ref: [Fe 6]. See also Baker's *Sharpening* II, Acta. Arith. **24** (1973) pp. 33–36, and [Ba 1], p. 46. This carries an effective improvement of Liouville's theorem.

(e) *Bounds for the exponents*
Catalan equation: Tijdeman's method, together with the lower bound of [VdP–L 2] leads to

$$y^q < x^p < \exp \exp \exp \exp 1000$$

for $x^p - y^q = 1$. (Michel Langevin has worked out this computation.)

$y^q = f(x)$, $f \in \mathbf{Q}[x]$ with at least two distinct roots, and $y > 1$, $x, y, q \in \mathbf{Z}$. Then

$$q \leqslant C \quad \text{where } C = C(f)\,.$$

Ref: Schinzel and Tijdeman, *On the equation* $y^m = P(x)$, Acta Arithm. **31** (1976) pp. 199–204.

Part II

Approximation of Logarithms

In this part we give the proofs for the diophantine inequalities needed for the bounds of integral points. These occur both for logarithms on the multiplicative group, and abelian logarithms (on elliptic curves).

It should be noted that the structure of the proofs involves the degree of the field over which one works. For our purposes, we regard this degree as fixed. There may be applications where one wishes to know the dependence on the degree, and also one may use the techniques of proof to give a lower bound for the degree of division points as in [L 10]. Results in this direction seemed too partial at the time this book is written to have included in the proofs explicitly the dependence on the degree. See for instance Baker [Ba 7] and Van der Poorten–Loxton [VdP–L 2], for the latest at time of appearance of this book. However, we wish to warn the reader to be aware of such possibilities.

As the theory is undergoing constant improvement, it is futile to attempt any exposition involving definitive results. I have merely tried to give typical results, meeting the following criteria:

1. The proofs in Chapters VIII and IX are entirely similar, thus emphasizing the analogy between the multiplicative group and elliptic curves. We advise the reader to read the proof of Chapter IX immediately after that of Chapter VIII, to absorb this analogy.

2. I have limited myself to statements which would be relatively easy to prove, and illustrate the analogy rather than go deeper in the first case. On the other hand I did not want to eliminate the Feldman polynomials, which one does not yet know how to use to improve the result in the elliptic case, to give the precise first power of $\log B$ in the inequality. Perhaps one has to use other interpolation polynomials.

3. The chapter involving Tijdeman's work on the Catalan equation is included first to show how one can bound degrees of equations instead of variables by means of the inequalities. It would be interesting to have extensions of this use to other contexts of algebraic geometry, although none are known at present.

Second, I wanted to illustrate other variations of the current methods. They stem from the work of Baker [Ba 2] and [Ba 3], who makes descents on heights and descents on systems of linear equations. As presented here, they reflect the improvements brought by Cijsouw-Waldschmidt, Tijdeman, and Van der Poorten.

4. Finally, the main purpose of this second part was to provide the proofs of the qualitative theorems in Chapter VI. I definitely did not intend this part as a full monograph on diophantine approximations of logarithms. The subject is undergoing such rapid changes with so many variations of the Baker method, and the

introduction of new methods (e.g. by Chudnovskii) that it is impossible at present to have an adequate perspective to set it in anything like definitive form. Especially the quantitative aspects provide an important area of research in several directions:

Giving an explicit determination of the constants involved, in which the geometry of numbers becomes essentially intermingled with algebraic number theory and the inequalities for linear combinations of logarithms. As mentioned before, these constants also exhibit a structure, going beyond making explicit whatever comes out of the current proofs, which is all that's happening at the moment. At their deepest, these constants connect with the Birch–Swinnerton–Dyer conjecture for elliptic curves.

Giving more accurate dependencies on the heights of the numbers, arising from the theory of successive minima.

Giving uniformities for the bounds of solutions, depending for instance only on the rank in the elliptic case.

Giving uniformities on the degrees of the number fields involved, thereby connecting with the theory of torsion points on elliptic curves, etc.

When the time is ripe, such quantitative results will themselves form a book. At that time, it will be useful for expository purposes to have available a simpler exposition for the qualitative results, which can serve as an introduction to the necessarily heavier machinery needed to exhibit the full structure of these more delicate invariants.

Chapter VII. Auxiliary Results

This chapter collects various lemmas of elementary algebraic number theory and complex variables, used to make estimates and to solve interpolation problems.

Although the height's formalism works very well in certain estimates, it is convenient in others to deal with bounds separately for the maximum of the conjugates of algebraic numbers and denominators. The first section makes explicit the relations between these different ways of estimating algebraic numbers.

§ 1. Heights and Sizes

Let K be a number field of degree d over $\mathbf{Q}$. If $\xi \in K$ we define

$\|\xi\|$ = maximum of the ordinary absolute values of the conjugates of ξ.

If Z is a set of algebraic numbers, we define a **denominator** for Z to be a positive integer a such that $a\xi$ is an algebraic integer for all $\xi \in Z$. We define

size of Z is $\leqslant B$ if $\|\xi\| \leqslant B$ for all $\xi \in Z$, and there exists a denominator a for all elements of z such that $a \leqslant B$.

It is easy to compare the size of a set and the heights of elements in that set. From the height property

$$H_K(\xi) = \prod_{v \in S_\infty} \max(1, |\xi|_v)^{n_v} \mathbf{N}\mathfrak{b},$$

where $\mathfrak{b}$ is a denominator ideal for ξ, we see at once that

$$\text{size } \xi \leqslant H_K(\xi).$$

On the other hand, if a is a denominator for ξ, then

$$H_K(\xi) = H_K(a\xi, a) = \prod_{v \in S_\infty} \max(|a\xi|_v, |a|_v)^{n_v}.$$

Thus we get another inequality

$$\boxed{\text{size } \xi \leqslant H_K(\xi) \leqslant (\text{size } \xi)^{2[K:\mathbf{Q}]} .}$$

On the other hand, since for any archimedean absolute value $|\ |$ we have

$$|\xi| \leqslant H_K(\xi) ,$$

and since $H_K(\xi) = H_K(1/\xi)$ for $\xi \neq 0$, we see that in this case we get

$$\boxed{\frac{1}{H_K(\xi)} \leqslant |\xi| \quad \text{whence} \quad \frac{1}{(\text{size } \xi)^{2[K:\mathbf{Q}]}} \leqslant |\xi| .}$$

This will be called the **fundamental arithmetic inequality**, or the **Liouville inequality**, for algebraic numbers. It gives a simple lower bound for the ordinary absolute value of an algebraic number. All the proofs in this part will consist in getting an upper bound in various situations, by analytic means, to contradict this Liouville inequality.

Let $\{w_1, \ldots, w_d\}$ be a basis of K over $\mathbf{Q}$. We have a multiplication rule

$$w_\nu w_\mu = \sum b_{\nu\mu\lambda} w_\lambda ,$$

where $b_{\nu\mu\lambda} \in \mathbf{Q}$, and are easily estimated in terms of the sizes of the basis. In fact, for any algebraic number $a \in K$ multiplication of basis elements by a is represented by a rational matrix whose coefficients are estimated as follows. We use the dual basis $\{w'_1, \ldots, w'_d\}$ such that

$$\operatorname{Tr}(w_\nu w'_\mu) = \delta_{\nu\mu} .$$

Let

$$\text{size}(w_1, \ldots, w_d) \leqslant W .$$

We can solve for the dual basis in terms of the basis by a system of non-homogeneous linear equations in K whose coefficients have a determinant which is

$$\det(\sigma_\nu w_\mu) .$$

where $\sigma_1, \ldots, \sigma_d$ are the conjugate embeddings of K in $\mathbf{C}$. Each w_i' is then a quotient of two determinants.

$$w_i' = \frac{D_i}{\det(\sigma_\nu w_\mu)}, \quad \text{where } D_i \text{ is a } d \times d \text{ determinant}.$$

For any absolute v on K we then find

$$|w_i'|_v \leqslant d!W^d |\det(\sigma_\nu w_\mu)^{-1}|_v .$$

Let a be a denominator (integer >0) for $(w_1, \ldots, w_d)$ and $a \leqslant W$. Then a^{2d} is a denominator for $\det(w_\nu w_\mu)^2$, which is a rational number, and a^d is a denominator for $\det(\sigma_\nu w_\mu)$. This yields

(*) $$|w_i'|_v \leqslant d!W^{2d}.$$

On the other hand, $|\det(\sigma_\nu w_\mu)|_v \leqslant d!\, W^d$, for every archimedean absolute value v. If b is a denominator for an algebraic number α in K then $b^d\mathbf{N}\alpha = \mathbf{N}\beta$ for the algebraic integer $\beta = b\alpha$, and $\mathbf{N}\beta$ is a denominator for α^{-1}. Consequently a denominator for $\det(\sigma_\nu w_\mu)^{-1}$ is bounded by $W^{d^2}(d!\, W^d)^d = d!^d W^{2d^2}$. Since a^d is a denominator for each determinant D_i, we conclude:

(**) $(w_1', \ldots, w_d')$ has a denominator bounded by $d!^d W^{2d^2+d}$.

Combining (*) and (**) yields:

Lemma 1. *If* size $(w_1, \ldots, w_d) \leqslant W$ *then*

$$\text{size}\,(w_1' \ldots, w_d') \leqslant d!^d W^{2d^2+d}.$$

Furthermore, let $\alpha \in K$. Then

$$\alpha w_\nu = \sum c_{\nu\lambda} w_\lambda$$

for suitable rational numbers $c_{\nu\lambda}$. Multiplying by w_μ' and taking the trace gives $c_{\nu\mu}$ on the right-hand side, and allows for an obvious estimate on the left-hand side in terms of the sizes of α, w_ν, w_μ'. Since $c_{\nu\lambda} = \operatorname{Tr}(\alpha w_\nu w_\lambda')$ we obtain:

Lemma 2. *Let* size $\alpha \leqslant A$, size $(w_1, \ldots, w_d) \leqslant W$. *Let Γ be the matrix $(c_{\nu\lambda})$. Then*

$$\text{size}\,\Gamma \leqslant dAWd!^d W^{2d^2+d+1}$$

Proof. Use (*) and (**) to estimate absolute values and denominators.

Remark. In the applications, we are frequently given a set of generators,

$$K = \mathbf{Q}(\beta_1, \ldots, \beta_N),$$

so that a basis for K can be extracted from monomials

$$\beta_1^{m_1} \ldots \beta_N^{m_N},$$

where the exponents m_j are bounded by the degrees of β_j. Thus we get an estimate for a basis of K in terms of the heights of the β_j, and their degrees.

§ 2. Linear Equations

We wish to give bounds for solutions of linear equations with integer coefficients, or coefficients in a number field, in terms of the height of these coefficients. The basic result giving such bounds is known as Siegel's lemma.

Lemma 1. *Let $n > r$. Let $A \leqslant 1$. Let*

$$\begin{aligned} a_{11}x_1 + \cdots + a_{1n}x_n &= 0 \\ \vdots \qquad\qquad & \vdots \\ a_{rl}x_l + \cdots + a_{rn}x_n &= 0 \end{aligned}$$

be a system of equations with integer coefficients a_{ij}, satisfying the bound $|a_{ij}| \leqslant A$. Then there exists a non-trivial solution X satisfying

$$|x_j| \leqslant 2(4nA)^{r/(n-r)}.$$

Proof. For any positive integer B we let

$$\mathbf{Z}^r(B) = \text{set of } Y \in \mathbf{Z}^r \quad \text{with } |Y| \leqslant B,$$

where $|\ |$ is the sup norm, if $Y = (y_1, \ldots, y_n)$ then

$$|Y| = \max |y_j|.$$

Let L be the linear map represented by the matrix (a_{ij}), so that we are seeking a solution of $LX = O$. Note that L maps $\mathbf{Z}^n(B)$ into $\mathbf{Z}^r(nAB)$. We want to select an integer B just so large that

$$\# \mathbf{Z}^n(B) > \# \mathbf{Z}^r(nAB).$$

We may then select $Y \neq Z$ in $\mathbf{Z}^n(B)$ such that $L(Y) = L(Z)$, and we let $X = Y - Z$

solve our problem. We have

$$\mathbf{Z}^n(B) \geqslant (2B+1)^n > (2B)^n$$

$$\mathbf{Z}^r(nAB) \leqslant (2nAB+1)^r \leqslant (4nAB)^r .$$

Thus we need $(2B)^n > (4nA)^r B^r$, from which the assertion of the lemma follows.

Lemma 2. *Let $n > rd$, and $A \geqslant 1$. Let $[K:\mathbf{Q}] = d$. Let*

$$\begin{aligned} \alpha_{11}x_1 + \cdots + \alpha_{1n}x_n &= 0 \\ \vdots \qquad\qquad & \\ \alpha_{r1}x_1 + \cdots + \alpha_{rn}x_n &= 0 \end{aligned}$$

be a system of linear equations, with coefficient matrix $\alpha = (\alpha_{ij})$ in K. Suppose that size $\alpha \leqslant A$. *Suppose K has a basis $\{w_1, \ldots, w_d\}$ whose size is bounded by W. Then there exists a non-trivial solution X in $\mathbf{Z}^n$ (ordinary integers!) satisfying the bound*

$$|X| \leqslant C_1(C_2 nAW^c)^{dr/(n-dr)}$$

where c, C_1, C_2 are numbers depending only on d, and easily determinable.

Proof. We write each α_{ij} in terms of the basis with rational coefficients. Then the system of linear equations is equivalent with another system

$$\begin{aligned} L_{11}(X)w_1 + \cdots + L_{1d}(X)w_d &= 0 \\ \vdots \qquad\qquad & \\ L_{r1}(X)w_1 + \cdots + L_{rd}(X)w_d &= 0 , \end{aligned}$$

where $L_{i\nu}$ is a linear form with rational coefficients, and we may clear denominators using the estimate of the lemma in § 1. Thus we obtain a system

$$L_{i\nu}(X) = 0, \quad i = 1, \ldots, r;\ \nu = 1, \ldots, d:$$

of rd equations in n unknowns to which Lemma 1 can be applied to conclude the proof.

By being very careful with the estimates, and the box principle, it is possible to give a very refined form of the Siegel lemma. Apparently the finest known is due to Mignotte, see Lemme 1.3.1 of Waldschmidt [Wa 1]. We state the result for the convenience of the reader, even though we shall not use it.

In the system of linear equations of Lemma 2, suppose that $\sigma_1, \ldots, \sigma_d$ are the embeddings of K in **C**, *and that α_{ij} are algebraic integers in K. Let A be a positive integer with*

$$A \geqslant \max_{i,\nu} \sum_{j=1}^{n} |\sigma_\nu(\alpha_{ij})| .$$

If $n > rd$ then the system has a non-trivial solution in $\mathbf{Z}^n$ satisfying

$$|x_j| \leqslant (\sqrt{2} \cdot A)^{dr/(n-dr)}.$$

§ 3. Estimates for Derivatives

Let P be a polynomial (in several variables) with coefficients in a number field K. By the **size of** P we shall mean the size of the set of its coefficients. We let $\|P\|$ denote the maximum of the absolute values of the conjugates of its coefficients.

Let

$$P(T_1, \ldots, T_n) = \sum \alpha_{i_1 \ldots i_n} T_1^{i_1} \ldots T_n^{i_n}$$

be a polynomial with complex coefficients, and let

$$Q(T_1, \ldots, T_n) = \sum \beta_{i_1 \ldots i_n} T_1^{i_1} \ldots T_n^{i_n}$$

be a polynomial with real coefficients $\geqslant 0$. We say that Q **dominates** P, and write $P \prec Q$, if $|\alpha_{(i)}| \leqslant \beta_{(i)}$ for all $(i) = (i_1, \ldots, i_n)$. It is then immediately verified that the relation of domination is preserved under addition, multiplication, and taking partial derivatives with respect to the variables $T_1, \ldots, T_n$. Thus if $P \prec Q$ then $\partial_i P \prec \partial_i Q$, where $\partial_i = \partial / \partial T_i$.

Lemma. *Let $f_1, \ldots, f_n$ be meromorphic functions in several variables $z_1, \ldots, z_r$, and assume that the partial derivatives $D_j = \partial / \partial z_j$ map the ring*

$$K[f_1, \ldots, f_n]$$

into itself. There exists a number C_1 having the following property. If $Q(T_1, \ldots, T_n) \in K[T_1, \ldots, T_n]$ is a polynomial with total degree $\leqslant L$, and

$$D^{(m)} = D_1^{m_1} \ldots D_r^{m_r}$$

is a differential operator of order $M = \sum m_j$, then

$$D^{(m)}(Q(f_1, \ldots, f_n)) = Q_{(m)}(f_1, \ldots, f_n),$$

where $Q_{(m)} \in K[T_1, \ldots, T_n]$ is a polynomial satisfying:

(i) $\deg Q_{(m)} \leqslant C_1(M + L)$

(ii) $\|Q_{(m)}\| \leqslant \|Q\| M! C_1^{M+L}$

(iii) *There exists a denominator for the coefficients of $Q_{(m)}$ bounded by* $\operatorname{den}(Q) C_1^{M+L}$.

Proof. For simplicity, we assume $r = 1$. Otherwise, the proof is the same with just some more indices. Let $P_j(T_1, \ldots, T_n)$ be a polynomial such that

$$\partial f_j/\partial z = P_j(f_1, \ldots, f_n) .$$

Let d be the maximum of the degrees of $P_1, \ldots, P_n$. There exists a differentiation $\bar{D}$ on the polynomial ring $K[T_1, \ldots, T_n]$ such that

$$\bar{D}T_j = P_j(T_1, \ldots, T_n) ,$$

and for any polynomial P we have

$$\bar{D}(P(T_1, \ldots, T_n)) = \sum_{j=1}^{n} (\partial_j P)(T_1, \ldots, T_n) P_j(T_1, \ldots, T_n) .$$

This is just obtained by the usual chain rule for differentiation, and $\partial_j = \partial/\partial T_j$. But the polynomial Q is dominated by

$$Q \prec \|Q\|(1 + T_1 + \cdots + T_n)^L$$

and each polynomial P_j is dominated by $\|P_j\|(1 + T_1 + \cdots + T_n)^d$. Thus for some constant C_2 we have

$$\bar{D}Q \prec \|Q\| C_2 L(1 + T_1 + \cdots + T_n)^{L+d} .$$

Proceeding inductively, we see that $\bar{D}^k Q$ is dominated by

$$\bar{D}^k Q \prec \|Q\| C_3^k L(L + d) \ldots (L + kd)(1 + T_1 + \cdots + T_n)^{L+kd} .$$

Since

$$L(L + d) \ldots (L + kd) \leqslant L(dL + d) \ldots (dL + kd) \leqslant d^k L(L + 1) \ldots (L + k) ,$$

this product is estimated by

$$d^k \frac{(L + k)!}{L!k!} Lk! \leqslant C_4^{L+k} k! .$$

This proves (i) and (ii) of the lemma. The third part is even easier.

Observe that the degrees of the polynomials $Q_{(m)}$ go up by an arithmetic progression.

In the applications, we want to evaluate a derivative

$$Df = Q_D(f_1, \ldots, f_n) , \quad D = D^{(m)} ,$$

at some point w, where $f = Q(f_1, \ldots, f_n)$ is a polynomial in the functions $f_1, \ldots, f_n$. Then all we have to do is plug in $f_1(w), \ldots, f_n(w)$ in $Q_D(T_1, \ldots, T_n)$ to obtain

$$Df(w) = Q_D(f_1(w), \ldots, f_n(w)).$$

If we have estimates for $|f_i(w)|$, then the lemma immediately implies estimates for $|Df(w)|$.

§ 4. Feldman Polynomials

Feldman [Fe 4] used binomial-type polynomials in his approximating function for improved estimates for linear forms in logarithms. His estimates were further extended by Baker and Tijdeman, cf. [Ba 3], [Ti 1]. We follow Baker (slightly refined by Tijdeman), who estimated the derivatives of Feldman polynomials.

Lemma 1. *Let $a_1, \ldots, a_s$ be arbitrary numbers, and let*

$$G(x) = (x - a_1)\ldots(x - a_s).$$

Let $D = d/dx$. Then

$$\frac{1}{m!} D^m G(x) = G(x) \sum \frac{1}{(x - a_{j_1})\ldots(x - a_{j_m})}$$

where the sum is taken over all choices of $(j_1, \ldots, j_m)$ among $\{1, \ldots, s\}$.

Proof. Induction, left to the reader.

In the applications, the numbers $a_1, \ldots, a_s$ will be consecutive integers, repeated with multiplicities.

For any positive integer k, we let

$$\nu(k) = \text{l.c.m. } 1, \ldots, k.$$

Then we have the estimate

(1)
$$\boxed{\nu(k) \leqslant e^{4k/3}.}$$

Indeed,

$$\nu(k) = \prod_{p \leqslant k} p^{[\log k/\log p]} \leqslant \prod_{p \leqslant k} k = k^{\pi(k)},$$

and the estimate follows from $\pi(k) \leqslant 4k/3 \log k$. In practice, any fixed constant in the exponent instead of 4/3 would do, and any simple estimate for prime numbers gives this, no need for the prime number theorem.

Let the **Feldman (binomial) polynomials** be

$$\Delta(x, k) = \frac{(x+1)\dots(x+k)}{k!}.$$

We apply Lemma 1 to the polynomial

$$G(x) = \Delta(x, k)^l (k!)^l.$$

Observe that in Lemma 1, the numbers $a_1, \dots, a_s$ need not be distinct, and are not distinct in the current applications. They are counted with multiplicities. *If x is any real number $\geqslant 1/2$, then we have an estimate*

$$(2) \qquad \boxed{\frac{1}{m!} D^m(\Delta(x, k)^l) \leqslant 4^{(x+k)l},}$$

because using the lemma, estimating each term of the sum by 1, we find:

$$\frac{1}{m!} D^m \Delta(x, k)^l \leqslant \binom{[x]+1+k}{k}^l \binom{lk}{m}$$

$$\leqslant 2^{([x]+1+k)l} 2^{kl} \leqslant 4^{(x+k)l}.$$

We also want an estimate for the denominator when x is a rational number, say

$$x = a/d$$

in lowest form. We have:

$$(3) \qquad \boxed{\operatorname{den} \frac{1}{m!} D^m(\Delta(a/d, k)^l \leqslant d^{2kl} \nu(k)^m \leqslant d^{2kl} e^{4km/3}.}$$

Proof. By Lemma 1,

$$d^{kl} \frac{1}{m!} D^m \Delta(x, k)^l = d^{kl} \left(\frac{(a/d+1)\dots(a/d+k)}{k!} \right)^l \sum \frac{1}{(a/d+j_1)\dots(a/d+j_m)}$$

$$= \left(\frac{(a+d)\dots(a+dk)}{k!} \right)^l d^m \sum \frac{1}{(a+dj_1)\dots(a+dj_m)}$$

We then estimate the powers of primes p occurring in the denominator.

Case 1. $p \nmid d$. Then the denominator contribution is at most the power of p in $(k!)^l$, which is

$$l\left(\left[\frac{k}{p}\right]+\cdots+\left[\frac{k}{p^t}\right]\right)$$

where $t = [\log k/\log p]$, but we have other cancellations which we shall now estimate to give a better bound. We write the expression we have to estimate as a sum of terms

$$\frac{((a+d)(a+2d)\dots(a+kd))^l}{k!(a+dj_1)\dots(a+dj_m)}.$$

Looking at $a+d, a+2d, \dots, a+kd \bmod p$, $\bmod p^2, \dots$ we see that the product $(a+d)\dots(a+kd)$ contains p at least

$$\left[\frac{k}{p}\right]+\cdots+\left[\frac{k}{p^t}\right]$$

times. Hence

$$\frac{((a+d)(a+2d)\dots(a+kd))^l}{(a+j_1 d)\dots(a+j_m d)}$$

contains p at least

$$l\left(\left[\frac{k}{p}\right]+\cdots+\left[\frac{k}{p^t}\right]\right)-mt$$

times. Hence the denominator contains p at most mt times, and therefore the denominator at p divides $v(k)_p^m$ (the maximal p-power in $v(k)^m$). This settles the present case.

Case 2. $p \mid d$. Then the terms in the sum do not have p in the denominators, and therefore the order of the p-contribution to the denominator is that of $k!$, which is $\leqslant kl$ because

$$\left[\frac{k}{p}\right]+\cdots+\left[\frac{k}{p^t}\right]\leqslant k.$$

In the present case, $p^{kl} \mid d^{kl}$, so the estimate is proved.

For want of a better place, we state here one more lemma on polynomials, which will be applied to powers of the Feldman polynomials.

Lemma 2. *Let P be a polynomial of degree $d \geqslant 1$. Let $0 \leqslant s \leqslant d$. Then the polynomials $P(x), P(x+1), \ldots, P(x+s)$ and $1, x, \ldots, x^{d-s-1}$ are linearly independent.*

Proof. By induction. We assume the lemma for degree d, and suppose P has degree $d+1$. Suppose that a linear combination

$$c_0 P(x) + c_1 P(x+1) + \cdots + c_s P(x+s)$$

has degree $\leqslant d-s$. We write this combination in the form

$$\begin{aligned} &c_0(P(x) - P(x+1)) + (c_0 + c_1)(P(x+1) - P(x+2)) + \cdots \\ &\qquad + (c_0 + \cdots + c_s)(P(x+s) - P(x+s+1)) + (c_0 + \cdots + c_s)P(x+s+1). \end{aligned}$$

Let $Q(x) = P(x) - P(x+1)$. Then Q has degree $\leqslant d$. Since $P(x+s+1)$ has degree $d+1$, it follows that

$$c_0 + \cdots + c_s = 0.$$

Applying induction to the polynomial $Q(x)$ shows that the other coefficients (partial sums of the c_j) are 0, whence $c_j = 0$ for $j = 1, \ldots, s$ as was to be shown.

§ 5. Estimates for Entire Functions

We recall elementary facts from complex function theory, to the effect that if an entire function has lots of zeros, then its maximum modulus is decreased accordingly.

Lemma 1. *Let f be holomorphic on the closed disc of radius R. Let $z_1, \ldots, z_S$ be distinct points inside the disc where f has zeros of multiplicities $\geqslant M$, and assume that these points lie in the disc of radius R_1, with*

$$R_1 \leqslant R/2.$$

Let $R_1 < R_2 \leqslant R$. Then on the circle of radius R_2 we have the estimate

$$\|f\|_{R_2} \leqslant \frac{\|f\|_R 4^{MS}}{(R/R_2)^{MS}}.$$

Proof. Let $|w| = R_2$. We estimate the function

$$\frac{f(z)}{((z - z_1)\dots(z - z_S))^M}((w - z_1)\dots(w - z_S))^M$$

on the circle of radius R. This function has precisely the value $f(w)$ at $z = w$. The estimate $|w - z_j| \leqq 2R_2$ is trivial, and the theorem follows at once.

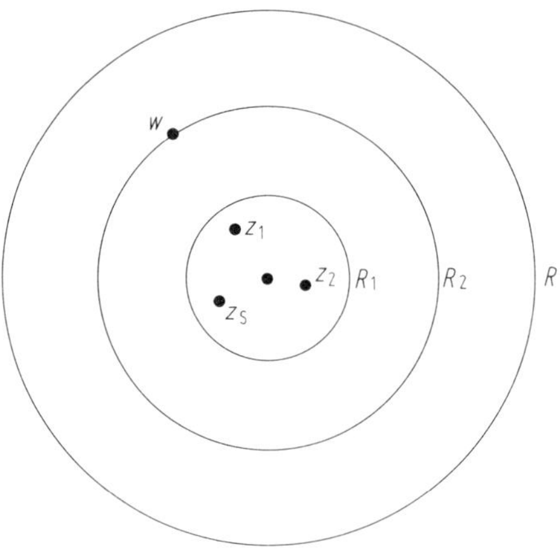

Fig. 7

If a function does not have zeros at certain points, but has small derivatives, then it is still true that the function is small in a region not too far away from these points. A quantitative estimate can be given, with a main term which is the same as if the function had zeros, and an error term, measured in terms of the derivatives. One technique (Siu's Lemma), usable in the theory of several variables, is to construct a polynomial of not too high degree having the same Taylor expansion (derivatives) at the given points, up to the given order, and subtracting it from the function to yield a new function with the appropriate zeros. Cf. [L 11], Appendix 2, § 2, self contained.

An older technique in one variable is via Cauchy's theorem.

Hermite Interpolation Formula. *Let $z_1, \dots, z_S$ be distinct points in the open disc of radius R, and let*

$$Q(z) = [(z - z_1)\dots(z - z_S)]^M.$$

Let f be holomorphic on the closed disc of radius R. Let Γ be the circle of radius R, and let Γ_j be a circle around z_j, not containing z_k for $k \neq j$, and contained in *the interior of Γ. Then for z not equal to any z_j, we have*

$$\frac{f(z)}{Q(z)}=\frac{1}{2\pi i}\int_{\Gamma}\frac{f(\zeta)}{Q(\zeta)}\frac{d\zeta}{(\zeta-z)}-\frac{1}{2\pi i}\sum_{j=1}^{S}\sum_{m=0}^{M-1}\frac{D^m f(z_j)}{m!}\int_{\Gamma_j}\frac{(\zeta-z_j)^m}{Q(\zeta)}\frac{d\zeta}{(\zeta-z)}.$$

Proof. This is a direct consequence of the residue formula. We consider the integral

$$\int_{\Gamma}\frac{f(\zeta)}{Q(\zeta)}\frac{1}{(\zeta-z)}d\zeta .$$

The function inside the integral has a simple pole at $\zeta = z$ with residue $f(z)/Q(z)$. This gives the contribution on the left hand side of the formula. The integral is also equal to the sum of the integrals taken over the small circles around the points $z_1, \ldots, z_S, z$. To find the residue at z_j, we expand $f(\zeta)$ at z_j, say

$$f(\zeta)=a_0+a_1(\zeta-z_j)+\cdots+a_M(\zeta-z_j)^M+\cdots .$$

Looking at the quotient by $Q(\zeta)$ immediately determines the residue at z_j in terms of coefficients of the expansion, which are such that

$$a_m = D^m f(z_j)/m! .$$

The formula then drops out.

It is then easy to estimate $f(z)$. Multiplying by $Q(z)$ introduces the quotients

$$\frac{Q(z)}{Q(\zeta)}=\prod_{j=1}^{S}\left(\frac{z-z_j}{\zeta-z_j}\right)^M$$

which are trivially estimated. The denominator is small according as the radius of Γ_j is small. In applications, one tries to take the Γ_j of not too small radius, and this depends on the minimum distance between the points $z, z_1, \ldots, z_S$. It is a priori clear that if the points are close together, then the information that the function has small derivatives at these points is to a large extent redundant. This information is stronger the wider apart the points are. Making these estimates, the following result drops out.

Lemma 2. *Let f be holomorphic on the closed disc of radius R. Let $z_1, \ldots, z_S$ be distinct points in the disc of radius R_1. Assume*

$$2R_1 < R_2 \quad \textit{and} \quad 2R_2 < R .$$

Let σ be the minimum of 1, and the distance between any pair of distinct points *among $z_1, \ldots, z_S$. Then*

$$\|f\|_{R_2} \leqslant \frac{\|f\|_R C^{MS}}{(R/R_2)^{MS}} + (CR_2/\sigma)^{MS} \max_{m,\, j} \frac{|D^m f(z_j)|}{m!}$$

where C is an absolute constant.

An estimate for the derivatives of f can be then obtained from Cauchy's formula,

$$\frac{D^k f(z)}{k!} = \frac{1}{2\pi i} \int_{|\zeta| = R_2} \frac{f(\zeta)}{(\zeta - z)^{k+1}} d\zeta ,$$

from which we see that $\frac{1}{k!} \|D^k f\|_{R_2/2}$ is estimated by a similar expression, multiplied by

$$2^{k+1}/R_2^k .$$

We may summarize the estimate of Lemma 2 by saying that the first term is exactly the same as would arise if f had zeros at the points $z_1, \ldots, z_S$, and the second term is a correcting factor describing the extent to which those points differ from actual zeros. In practice, the derivatives of f are very small at these points, which thus do not differ too much from zeros.

In certain delicate applications, the set of points is chosen specially, for instance to consist of successive integers, and one can then improve one of the estimates to get rid of what would turn out to be a log log term arising from an $S!$.

Lemma 3. *Let z be a complex number $\neq j$ for $j = 1, \ldots, S$. Suppose that $|z| \leqslant BS$, with $B \geqslant 1$. Let δ be the smallest distance from z to the set of integers $1, \ldots, S$. Then*

$$\delta C^{-S} S! \leqslant |(z-1)\ldots(z-S)| \leqslant (BC)^S S!$$

where C is an absolute constant.

Proof. For the right inequality, we write

$$|z - j| \leqslant |z| + j \leqslant BS + Bj .$$

We factor out B, and the inequality drops out. For the left inequality, suppose that there is an index j such that

$$j < |z| < j + 1 .$$

If $1 < k < j$ then $|z - K| \geqslant |z| - k \geqslant j - k$. On the other hand if $j+1 \leqslant k$, then $|z-k| \geqslant k-j$. Taking the products and using the fact that the binomial coefficient is bounded by 2^S yields the lemma, when $|z|$ satisfies the stated bound. This is obviously the most delicate situation. If $|z|$ becomes larger, then the estimates improve, and we leave this to the reader.

If we now assume that $z_j = j$ for $j = 1, \ldots, S$ in the Hermite formula, then we can estimate the quotient, for $|z| = R_2$ and ζ on a circle of radius $\frac{1}{4}$ centered at one of the points:

$$\left|\frac{Q(z)}{Q(\zeta)}\right| = \left|\frac{(z-1)\ldots(z-S)}{(\zeta-1)\ldots(\zeta-S)}\right|^M \leqslant C^{MS}\left(\frac{R_2}{S}\right)^{MS},$$

which is an improvement over the previous estimate. Therefore we obtain:

Lemma 4. *If in Lemma* 2 *we have $z_j = j$ for $j = 1, \ldots, S$, then the estimate can be improved to:*

$$\|f\|_{R_2} \leqslant \textit{same first term} + (CR_2/S)^{MS} \max_{m,\, j} \frac{|D^m f(z_j)|}{m!}$$

§ 6. The p-Adic Case

Since the work of Mahler [Mah 1] it has been known that transcendence and approximation theorems in the complex case have analogues in the p-adic case. We shall not go into this deeply, but in this section merely make a few remarks on the foundations to help the reader get into the literature. In the past, known patterns of proof have usually gone over from the complex case to the p-adic case without major difficulties, although there are some significant exceptions to this, some of which are mentioned below. Cf. also [L 3], Appendix.

Let K be complete under a non-archimedean valuation. Let

$$f(z) = \sum_{n=0}^{\infty} a_n z^n$$

be a formal power series with coefficients in K. If $R > 0$ we define

$$|f|_R = \sup R^n |a_n| .$$

Then $|\ |_R$ is a valuation on the space of power series f for which $|f|_R$ is finite. For such f, the series $f(z)$ converges absolutely if $|z| < R$, and if $R^n|a_n| \to 0$ as $n \to \infty$, then the series also converges absolutely for $|z| \leqslant R$. We have in fact

$$|f(z)| \leqslant |f|_R .$$

As with Gauss' Lemma, one can prove that for two power series we have

$$|fg|_R = |f|_R|g|_R .$$

We leave this as an exercise to the reader. It will also follow from the next remark.

Because the value group may not be dense in the positive reals, for instance, one cannot for arbitrary K define $|f|_R$ in the usual manner as a sup over the disc of radius R. However, one can always embed K in the completion of its algebraic closure, and then the expected characterization is true, as stated in the next lemma.

Lemma. *Assume that K is complete and algebraically closed. Then*

$$|f|_R = \sup_{|z|=R} |f(z)| .$$

Let $\{b_m\}$, $m \in M$, be those coefficients such that $|b_m|R^m$ is maximal. By considering a radius $R' < R$ and letting $R' \to R$, we can reduce the proof to the case when $R^n|a_n| \to 0$ as $n \to \infty$, so there is only a finite number of such indices m. We investigate the polynomial

$$P(z) = \sum_{m \in M} b_m z^m .$$

Let z be a fixed element of K with $|z|$ close to R, and let u range over units. We have

$$P(z) = \sum b_m u^m z^m = \sum c_m u^m .$$

We have to select u such that

$$\begin{aligned} |\textstyle\sum c_m u^m| &= |c_k u^k| \quad \text{for some } k \\ &= |c_k| . \end{aligned}$$

We pick k such that $|c_k|$ is maximal among $|c_m|$, and write

$$P(z) = c_k Q(z) ,$$

where Q has integral coefficients, and the k-th coefficient is 1. We can then select u to be a unit such that $Q(u) \not\equiv 0 \bmod \mathfrak{p}$, where $\mathfrak{p}$ is the maximal ideal of the valuation. This can be done if the residue class field is algebraically closed, and concludes the proof.

The crucial analytic estimates depended on a Jensen–Schwarz inequality. In the p-adic case, it is due to Mahler.

Theorem 6.1. *Let $0 < R_1 < R$. Suppose f has N roots (counting multiplicities) in the disc $|z| \leqslant R_1$. Assume $|f|_R$ finite. Then*

$$|f|_{R_1} \leqslant (R_1/R)^N |f|_R .$$

Proof. Write

$$f = Pg$$

where

$$P(z) = (z - z_1)\dots(z - z_N)$$

and $|g|_R < \infty$. We have

$$|P|_R = R^N \quad \text{and} \quad |P|_{R_1} = R_1^N .$$

Hence

$$\begin{aligned} |f|_{R_1} = R_1^N |g|_{R_1} &\leqslant (R_1/R)^N R^N |g|_R \\ &\leqslant (R_1/R)^N |f|_R . \end{aligned}$$

This proves Mahler's theorem.

Serre [Se 1] has formulated and proved a higher dimensional version, under an additional hypothesis of equidistribution of zeros, and without multiplicities. These conditions were eliminated by Robba [Ro].

The analogue of the estimate for a function having small values follows either by subtracting an appropriate polynomial whose coefficients are easily estimated, or by using the p-adic analogue of the Cauchy–Hermite interpolation formula. A convenient account of p-adic analysis will be found in Serre [Se 1], see also Adams [A], whose work was further generalized by Bertrand [Be 1], [Be 2], [Be 3], [Be 4], in a manner applicable to the Weierstrass elliptic functions, and also the Jacobi–Tate functions on the multiplicative model of an elliptic curve. See Coates for ordinary logarithms.

Introduction to the Baker Method

For the reader unacquainted with the techniques of diophantine approximation in transcendence proofs, it will be useful to see these techniques applied in a simple case (actually the first case considered by Baker after the work of Gelfond). Some of the technical complications required to deal with the refined inequalities disappear, and the structure of the proof becomes more accessible. Therefore we prove here the following statement.

Let $\alpha_1, \ldots, \alpha_{r+1}$ *be algebraic number which are multiplicatively independent, i.e. there is no relation* $\prod \alpha_j^{n_j} = 1$ *with integers* n_j *unless* $n_j = 0$ *for all* j. *Then* $\log \alpha_1, \ldots, \log \alpha_{r+1}$ *are linearly independent over the algebraic numbers.*

Proof. Suppose we have a linear dependence

$$\beta_1 u_1 + \cdots + \beta_r u_r - u_{r+1} = 0$$

where β_j are algebraic, say in K, and not all 0, and

$$u_j = \log \alpha_j$$

with any fixed determination of the log, say the principal value. We also assume that K contains all the α_j. We shall then prove:

There exists a polynomial

$$P(X_1, \ldots, X_{r+1}) = \sum a_{(\lambda)} X_1^{\lambda_1} \ldots X_{r+1}^{\lambda_{r+1}}$$

with integer coefficients $a_{(\lambda)}$ *not all* 0, *of degree* $\leqslant L$ *in each variable, with* L *arbitrarily large, such that*

$$P(\alpha_1^{1/q}, \ldots, \alpha_{r+1}^{1/q}) = 0$$

for some prime number $q > 2L$.

From this main lemma, we obtain a contradiction as follows. Let

$$K_s = K(\alpha_1^{1/q}, \ldots, \alpha_s^{1/q}) .$$

Then for q sufficiently large,

$$[K_{s+1} : K_s] = q \, .$$

By the main lemma, it follows that the polynomial

$$P(\alpha_1^{1/q}, \ldots, \alpha_r^{1/q}, X_{r+1}) \text{ of degree } \leqslant L$$

is identically 0, because $\alpha_{r+1}^{1/q}$ is a root. Recursively, we can go down the variables to conclude that P itself is identically 0, whence the contradiction that not all $a_{(\lambda)}$ are 0.

Step 1. To construct the polynomial P, we construct the auxiliary function

$$F(z_1, \ldots, z_r) = \sum a_{(\lambda)} \, e^{\lambda_1 z_1} \ldots e^{\lambda_r z_r} \, e^{(\beta_1 z_1 + \cdots + \beta_r z_r)\lambda_{r+1}}$$

with $0 \leqslant \lambda_j \leqslant L - 1$ and integer coefficients $a_{(\lambda)}$ to have zeros of high order at the points $(nu_1, \ldots, nu_r)$ for sufficiently many positive integers n. More precisely, let

$$D^{(m)} = D_1^{m_1} \ldots D_r^{m_r}$$

be a differential operator of order $|m| = m_1 + \cdots + m_r$. We require that

$$D^{(m)} F(nu_1, \ldots, nu_r) = 0 \quad \text{for } |m| \leqslant M, \; 1 \leqslant n \leqslant N$$

where L, M, N are parameters which are still to be chosen. The vanishing of the above derivatives amounts to a system of linear equations in K, whose coefficients are easily determined and involve the monomials

$$(\alpha_1^{\lambda_1} \ldots \alpha_r^{\lambda_r})^n \, .$$

We want to apply Siegel's lemma. For this purpose, we make the number of variables approximately equal to the number of equations, but slightly larger. We have:

Number of variables $= L^{r+1}$
Number of equations $= M^r N$
Size of coefficients $\leqslant C^{(L+M)N}$,

where C is a constant depending on K, α_j, β_j. We let

$$L^{r+1} = c_0 M^r N, \qquad N = [\log M] \text{ (say)}$$

where c_0 is some constant $\geqslant 2$. It does not matter much what we pick for c_0. Taking M, N appropriately divisible so that $2M^r N$ comes out an exact $(r + 1)$-th power, we could take $c_0 = 2$. Then we can can solve for the $a_{(\lambda)}$ in $\mathbf{Z}$, not all 0, satisfying a similar upper bound in absolute value as the size of the coefficients of the linear equations.

Step 2. The next step is an extrapolation on integral multiples. At the cost of diminishing the order of derivatives only slightly, we expand greatly the range of n where these derivatives vanish. Precisely:

Let $b \geqslant 1$ be fixed. There exists $0 < \delta < 1$ having the following property. Let v be an integer $\geqslant 0$, and let

$$S_v = NM^{v\delta}.$$

Suppose $v\delta \leqslant b$. If $|m| \leqslant M/2^v$ and $n \leqslant S_v$, then

$$D^{(m)}F(nu_1, \ldots, bu_r) = 0.$$

Proof. By induction on v. For $v = 0$ this is a construction. We assume the result for v. Let

$$|m'| \leqslant M/2^{v+1},$$

and let

$$g(z) = D^{(m')}F(zu_1, \ldots, zu_r),$$

so that g is a function of one variable. Then for $k \leqslant M/2^{v+1}$ the k-th derivative $D^k g(z)$ is a simple linear combination of derivatives

$$D^{(m')+(\rho)}F(zu_1, \ldots, zu_r),$$

which vanish at $z = n$ for $n \leqslant S_v$, and we can therefore apply the standard estimates from elementary complex variables to estimate g. Let $n' \leqslant S_{v+1}$. We select a circle of radius

$$R = NM^{(v+1)\delta}M^{\varepsilon} = S_{v+1}M^{\varepsilon} \quad \text{with small } \varepsilon.$$

The term M^{ε} is put here to tend to infinity. Then

$$|g(n')| \leqslant \|g\|_{S_{v+1}} \leqslant \frac{\|g\|_R C_2^{MS_v}}{(R/S_{v+1})^{MS_v}}.$$

As $M \to \infty$ we want the right hand side to tend to 0, at the same rate as the denominator. Let us agree that if X, Y are two quantities, with

$$X = M^x \quad \text{and} \quad Y = M^y$$

then we write $X \prec Y$ if $x < y$. We have

$$\|g\|_R \leqslant C^{LR} = C^{LS_{v+1}M^{\varepsilon}}.$$

All we need to do is pick δ, ε small enough that

$$LS_{\nu+1}M^{\varepsilon} < MS_{\nu}.$$

This is certainly possible since $S_{\nu+1}/S_{\nu} = M^{\delta}$, and L is at most a power of M strictly less than 1. Thus we conclude that

$$(*) \qquad |g(n')| \leqslant C_3^{-MS_{\nu}}.$$

On the other hand $g(n')$ is an algebraic number in K, whose size is easily estimated, and in fact,

$$(**) \qquad \text{size } g(n') \leqslant C_4^{MN} C_5^{LS_{\nu+1}}.$$

If $g(n') \neq 0$, the estimates (*) and (**) are contradictory, thus concluding the second step.

Step 3. We now prove that for a prime q with $2L \leqslant q \leqslant 4L$ we have

$$F\left(\frac{1}{q}u_1, \ldots, \frac{1}{q}u_r\right) = 0.$$

This will conclude the proof of the main lemma.

We let

$$g(z) = F(u_1 z, \ldots, u_r z).$$

Repeating the standard estimate as in Step 2, for the function which has many zeros, we find

$$\left|g\left(\frac{1}{q}\right)\right| \leqslant C_6^{-MS},$$

where S is the largest value of S_{ν} to which we pushed the inductive step. Furthermore, $g(1/q)$ is an algebraic number, which of course is not necessarily rational over K, and has degree $\ll q^r$. Its size is easily estimated by C_7^{MN}. If $g(1/q) \neq 0$, then we get the lower bound

$$C_8^{-MNq^r} \leqslant \left|g\left(\frac{1}{q}\right)\right|.$$

To reach a contradiction it suffices that

$$MNq^r < MS,$$

or in other words, that

$$MNL^r < NM^{\nu\delta}$$

for the largest value of ν in Step 2. Certainly the smallest ν such that

$$\nu\delta > r + 1$$

will suffice. It is of course trivial to have solved previously for δ, and now for such ν in terms of r. The main point here is that the inductive step needed to be pushed only to a point depending only on r, which could have been explicitly stated a priori. We preferred to develop the method, derive necessary conditions to see how it would go through, and arrive at the determination of δ, ν only at the end. In any case, this concludes the proof of the theorem.

All subsequent proofs will follow a similar pattern, but with added complications due to the following circumstances.

1. We do not merely wish to prove linear independence, we want to give a measure for it, that is, give a lower bound for the absolute value of the sum of logarithms. For this it is necessary to keep track of the heights of α_j, β_j throughout the proof. Depending on how much one is interested, one can also keep track of the degree of the field.

2. If we want the dependence on the height of β_j to have the value $\log B$ as in the Baker–Feldman theorem, then it is necessary to introduce an auxiliary function which in addition to the exponential terms will also contain polynomial coefficients, following techniques previously developed by Gelfond and Feldman. The system of linear equations then becomes more complicated.

3. For elliptic functions, which are quotients of entire functions of order 2 rather than entire functions of order 1 like the ordinary exponential function, some technical lemmas must be introduced to deal with the appearance of denominators.

4. For the refined statements of Baker–Tijdeman (or those of the last chapter), where we wish to improve the dependence on the heights of the numbers α_j, we introduce (following Baker again) descents whose purpose is to make some kind of induction on those heights.

5. Instead of dealing with multiplicatively independent numbers, we deal more generally with numbers whose logs are linearly independent over the rational numbers. This amounts merely to considering periods in addition to the logs of numbers not at the origin.

Chapter VIII. The Baker–Feldman Theorem

This chapter establishes a good inequality for linear combinations of ordinary logarithms of algebraic numbers. As is well-known, special cases were known to Gelfond for two logarithms, but Baker was the first to see how to deal with more than two. Feldman [Fe 4] then obtained an inequality which is especially good with respect to the heights of the coefficients of the relation, using special interpolation polynomials.

The proof we give is essentially that of Baker–Feldman, cf. for instance Baker's book [Ba 1]. However, it is convenient to make an a priori reduction to the case when the logarithms of algebraic numbers are linearly independent. This simplifies systematically some aspects of the proof, both in the present chapter and subsequent ones. Although the matter is simple, it is not generally understood (cf. for instance the misleading statement at the end of Chapter 3 of [Ba 1], where neither Lemma 3, Lemma 7, nor the inductive argument are needed).

Furthermore, as in Feldman, Stark [St 3], Baker–Stark [BS], and Baker's papers, we keep track of the dependence of the approximating function on the heights of all algebraic numbers under consideration, not just the coefficients. This is especially important for certain applications, as in the next chapter, allowing for inductive arguments on these heights. It is such uniformity which allows for uniformity theorems concerning integral diophantine equations as in Tijdeman's theorem.

§ 1. Statement of the Theorem

Let $\alpha_1, \ldots, \alpha_n$ be algebraic numbers in a number field K. We shall use the notation

$$u_j = \log \alpha_j$$

where the log is taken with principal value. We let

$$U_j = \log H_K(\alpha_j) \quad \text{and} \quad U = \log H_K(1, \alpha_1, \ldots, \alpha_n)\,.$$

In the applications we shall consider uniformity only with respect to the degree $[K:\mathbf{Q}]$ and n, and so we shall not pay special attention to distinguish the absolute from the relative height, although it would be easy to do so.

Throughout this chapter we let

$$\tau(B, U) = (\log B)U^{\kappa},$$

where $\kappa > (r + 3)^2 + 1$, and $r + 1$ is the maximum number of linearly independent elements among $u_1, \ldots, u_n, 2\pi i$.

Theorem 1.1. *Let* $\beta_1, \ldots, \beta_n \in K$ and assume

$$\beta_1 u_1 + \cdots + \beta_n u_n \neq 0.$$

Let $B = H_K(1, \beta_1, \ldots, \beta_n)$. *Then there is an inequality*

$$|\beta_1 u_1 + \cdots + \beta_n u_n| \geqq C_0^{-\tau(B, U)},$$

where C_0 *depends only on* n *and* $[K : \mathbf{Q}]$.

The rest of this chapter will provide the proof.

We shall first perform a reduction which allows us to assume that the u_j are linearly independent over the integers. We assume that Theorem 1.1 is false, and thus that we have the inequality

$$(1) \qquad \boxed{|\beta_1 u_1 + \cdots + \beta_n u_n| \leqq C_0^{-\tau(B, U)}.}$$

where C_0 is a sufficiently large constant.

First, we observe that without loss of generality, we may, and shall always assume that $u_1 = 2\pi i$.

Next, we select inductively

$$u_1, u_{j2}, \ldots, u_{jr+1}$$

with $1 < j_2, < \cdots < j_{r+1}$ to be maximal linearly independent among $u_1, \ldots, u_n$. We express the remaining u_j as linear combinations of these with rational coefficients, which have heights

$$\ll U^r \log\log U$$

by Chapter IV, Theorem 5.2. Substituting these linear combinations in the original linear combinations, and renumbering our elements if necessary, we get

$$\beta_1 u_1 + \cdots + \beta_n u_n = \beta'_1 u_1 + \cdots + \beta'_r u_r + \beta'_{r+1} u_{r+1}$$

and the coefficients β'_j satisfy

$$H_K(\beta'_j) \ll B \cdot U^r \log\log U.$$

Let $B' = \max H_K(\beta'_j)$. We have the inequality

$$(2)\qquad |\beta'_1 u_1 + \cdots + \beta'_{r+1} u_{r+1}| \leqslant C_0^{-(\log B)U^\kappa}.$$

We want the right hand side to be bounded by

$$C_1^{-(\log B')^\kappa}.$$

This would give the desired reduction. It suffices that

$$\log B' \ll \log B .$$

If $U^r \log\log U \leqslant B$ then this is certainly the case. If

$$B \leqslant U^r \log\log U$$

then

$$\log B' \ll \log U .$$

Consequently we get in this case

$$(3)\qquad |\beta'_1 u_1 + \cdots + \beta'_{r+1} u_{r+1}| \leqslant C_3^{-(\log B')U^\kappa/\log U}.$$

This again achieves the desired reduction.

We still perform one final step. We divide both sides by

$$|-\beta'_{r+1}| .$$

This does not change the shape of the right-hand side (we replace C_3 by another constant), and the left hand-side is further normalized in the form which is most convenient for the proof. We have then reduced Theorem 1.1 to the following statement.

Theorem 1.2. *Let $u_j = \log \alpha_j$ $(j = 1, \ldots, r+1)$ and let*

$$U = \log H_K(1, \alpha_1, \ldots, \alpha_{r+1}) .$$

Assume that $u_1, \ldots, u_{r+1}$ are linearly independent over the integers. Let $\beta_1, \ldots, \beta_r \in K$ and let $B = H_K(1, \beta_1, \ldots, \beta_r)$. Then $u_1, \ldots, u_{r+1}$ are linearly independent over the algebraic numbers, and we have

$$|\beta_1 u_1 + \cdots + \beta_r u_r - u_{r+1}| \geqslant C_0^{-\tau(B,U)}$$

where C_0 is a constant depending only on r and $[K:\mathbf{Q}]$.

§ 2. Main Lemma and its Application

For the rest of this chapter we shall use the following notation.

$M = M_0 U^\sigma$ where M_0 is a sufficiently large constant, and $\sigma > r$. For the proofs of the next sections, it will be necessary to take σ a little larger, $\sigma = r + 3$ will do.

$N = [\log(MB)]$

$L \gg\ll M^t$ where t is a number with $0 < t < 1$. It will turn out that

$$t = \frac{r+1}{r+2}.$$

Main Lemma. *Assume that we have the inequality*

$$|\beta_1 u_1 + \cdots + \beta_r u_r - u_{r+1}| < C_0^{-\tau(B,U)}$$

for a sufficiently large constant C_0. Then there exists a polynomial, not identically zero,

$$P(z_0, X_1, \ldots, X_{r+1}) = \sum\sum a_{(\lambda)} \Delta(z_0 + \lambda_{-1}, N)^{\lambda_0} X_1^{\lambda_1} \ldots X_{r+1}^{\lambda_{r+1}}$$

satisfying:

(i) $\quad 0 \leqslant \lambda_0, \ldots, \lambda_{r+1} \leqslant L$ *and* $0 \leqslant \lambda_{-1} \leqslant N$

(ii) *The coefficients $a_{(\lambda)}$ are integers.*

(iii) *We have*

$$P\left(\frac{n}{q}, \alpha_1^{n/q}, \ldots, \alpha_{r+1}^{n/q}\right) = 0$$

for every positive integers $q \leqslant M$ and $n \leqslant NM$.

In this last condition, by definition,

$$\alpha_j^{n/q} = e\left(\frac{n}{q} u_j\right) \quad \text{where} \quad e(z) = e^z.$$

In particular, we select a prime number $q \geqslant M/2$. Then by Kummer Theory, putting $K_s = K(\alpha_1^{n/q}, \ldots, \alpha_s^{n/q})$ we have

$$[K_s : K_{s-1}] = q - 1 \quad \text{if } s = 1 \quad \text{and } q \text{ if } s \geqslant 1,$$

for any positive integer n prime to q. For such n with

$$1 \leqslant n \leqslant MN$$

the polynomial

$$P\left(\frac{n}{q}, X_1, \ldots, X_{r+1}\right)$$

vanishes at $\alpha_1^{n/q}, \ldots, \alpha_{r+1}^{n/q}$ and is consequently identically zero, according to the following obvious lemma.

Lemma. *Let K be a field. Let $\xi_1, \ldots, \xi_n$ be algebraic over K. Put*

$$K_s = K(\xi_1, \ldots, \xi_s), \quad s = 1, \ldots, n$$

and let $[K_s : K_{s-1}] = d_s$. Let

$$P(X_1, \ldots, X_n) \in K[X_1, \ldots, X_n]$$

be a polynomial with coefficients in K, of degree $< d_j$ in each variable X_j. If $P(\xi_1, \ldots, \xi_n) = 0$ then P is identically zero.

This implies that each polynomial

$$\sum_{\lambda_{-1}, \lambda_0} a_{(\lambda)} \Delta(z_0 + \lambda_{-1}, N)^{\lambda_0}$$

for each choice $(\lambda_1, \ldots, \lambda_{r+1})$ vanishes at the points n/q. By the prime number theorem, there are

$$\gg \frac{NM}{\log\log M}$$

such fractions. However, each one of these polynomials has degree

$$\leqslant NL,$$

and cannot be identically zero by the linear independence of the Feldman polynomials, Lemma 2 of Chapter VII, § 4. Since L is essentially equal to a power of M less than 1, we get a contradiction which proves the theorem.

Observe that what is essential in the main lemma is that the polynomial should vanish at the irrationalities $\alpha_j^{n/q}$ for q large compared to the degree.

Using division values in connection with arguments of diophantine approximations is a technique used for the first time independently and simultaneously by Coates [Co 3] on elliptic curves, and Baker–Stark [BS] on the multiplicative group.

§ 3. Construction of the Approximating Function

Let $a_{(\lambda)}$ denote integers to be determined, and let

$$F(z_0, \ldots, z_r) = \sum a_{(\lambda)} \Delta(z_0 + \lambda_{-1}, N)^{\lambda_0} e^{z_1\lambda_1} \ldots e^{z_r\lambda_r} e^{(\beta_1 z_1 + \cdots + \beta_r z_r)\lambda_{r+1}}$$

where

$$0 \leqslant \lambda_0, \ldots, \lambda_{r+1} \leqslant L, \qquad 0 \leqslant \lambda_{-1} \leqslant N$$

and

$$N = [\log(MB)], \qquad M = M_0 U^\sigma$$

$$L^{r+2} = 4[K : \mathbf{Q}] M^{r+1} .$$

We treat σ as unknown and determine it so as to make the method of proof successful. Principally, it will have to satisfy the condition in § 5 that

$$\frac{r+1}{r+2} + \frac{1}{\sigma} < 1 ,$$

so $\sigma = r + 3$ will do. It turns out any other requirements on σ are subsumed by this one.

We note that F is equal to a polynomial

$$P(z_0, X_1, \ldots, X_{r+1})$$

evaluated at:

$$X_j = e^{z_j} \quad \text{for } j = 1, \ldots, r$$

$$X_{r+1} = e(\beta_1 z_1 + \cdots + \beta_r z_r) .$$

Let

$$D^{(m)} = D_0^{m_0} \ldots D_{r+1}^{m_{r+1}}$$

be a differential operator. Then

$$D^{(m)} F = P_{(m)}(z_0, e^{z_1}, \ldots, e^{z_r}, e^{(\beta_1 z_1 + \cdots + \beta_r z_r)})$$

is a polynomial. We shall use the notation

$$D^{(m)} F(z_0, z_1, \ldots, z_r; z_{r+1}) = P_{(m)}(z_0, e^{z_1}, \ldots, e^{z_r}, e^{z_{r+1}}) .$$

Thus we have substituted z_{r+1} for $\beta_1 z_1 + \cdots + \beta_r z_r$.

If n is an integer, then the *substituted* polynomial

$$D^{(m)}F(n, nu_1, \ldots, nu_r; nu_{r+1}) = P_{(m)}(n, \alpha_1^n, \ldots, \alpha_{r+1}^n)$$

is an algebraic number in K. (If instead of n being an integer, we replaced n by a rational number, then the substituted polynomial would have values in an algebraic extension of K.)

We require that

$$\frac{1}{m_0!} D^{(m)}F(n, nu_1, \ldots, nu_r; nu_{r+1}) = 0$$

for

$$|m| \leqslant M \quad \text{and} \quad 1 \leqslant n \leqslant N.$$

This amounts to a system of linear equations for the $a_{(\lambda)}$ of the type considered in Chapter VII, § 2, Lemma 2. We have

$$\text{Number of variables} \geqslant NL^{r+2}$$

$$\text{Number of equations} \leqslant NM^{r+1}.$$

The choice of L in relation to M is made exactly so that we can apply that lemma, and so that the Dirichlet exponent has a bounded value. We also have

$$\text{Size of coefficients} \leqslant C^{NLU} B^{cM} L^M C^{MN}$$

with appropriate constants C, c. The triviality of the differential equation satisfied by the ordinary exponential function makes even the considerations of Chapter VII, § 3 unnecessary, but we have to use the Feldman–Baker estimates of Chapter VII, § 4 for the Feldman polynomials, and their derivatives. The Siegel lemma then implies that we can solve for $a_{(\lambda)}$ in integers not all zero, satisfying the same bound as the above coefficients. We picked $\sigma \geqslant r + 2$ as an exponent for $M = M_0 U^\sigma$ so that

$$LU \leqslant M.$$

Consequently we get the estimate

$$|a_{(\lambda)}| \leqslant C^{MN}.$$

We are going to extrapolate the set of points where the substituted derivatives are equal to 0. For this purpose we need two estimates, showing that the actual derivative is very close to the substituted derivative within the desired range, and showing that the height does not increase too fast. The final extrapolation will show that the original polynomial P representing the function F,

$$F(z_0, \ldots, z_r \ z_{r+1}) = P(z_0, e^{z_1}, \ldots, e^{z_r}, e^{z_{r+1}}),$$

will satisfy the conditions of the main lemma.

It will be convenient to use the following notation. If X, Y are quantities which are expressed as powers of M or U, say $X = M^a X'$ and $Y = M^b Y'$ where X', Y' do not contain M (or U), we write

$$X \prec Y \quad \text{to mean} \quad a < b.$$

§ 4. Two Estimates

Lemma 1. *Let $b \geqslant r$. Let q be a positive integer $\leqslant M$. Let n be a positive integer with $n \leqslant S$.* If $S = NM^b$, *and*

$$SUL \prec (\log B) U^{\kappa}$$

then for $|m| \leqslant M$ we have for C_0 sufficiently large,

$$\left| \frac{1}{m_0!} D^{(m)} F\left(\frac{n}{q}, \frac{n}{q} u_1, \ldots, \frac{n}{q} u_r; \frac{n}{q} u_{r+1}\right) - \frac{1}{m_0!} D^{(m)} F\left(\frac{n}{q}, \frac{n}{q} u_1, \ldots, \frac{n}{q} u_r\right) \right|$$

$$\leqslant C_0^{-\tau(B, U)/4}.$$

Proof. The polynomial $P_{(m)}$ involves derivatives of the Feldman polynomials, and monomials in the exponentials. From Chapter VII, § 4 we know that

$$\text{size} \, \frac{1}{m_0!} D^{m_0} \Delta\left(\frac{n}{q}, N\right)^{\lambda_0} \leqslant C_5^{NL + NM + nL} \leqslant C_6^{NM + SL}.$$

The coefficients of the polynomial $P_{(m)}(n/q, X_1, \ldots, X_{r+1})$ are bounded by this last expression. We then have to estimate the difference of monomials

$$X_1^{\lambda_1} \ldots X_{r+1}^{\lambda_{r+1}}$$

evaluated at

$$X_j = e^{nu_j/q} \quad \text{for} \quad j = 1, \ldots, r$$

and

$$X_{r+1} = e^{w} \quad \text{or} \quad e^{w_0}$$

where

$$w = \frac{n}{q}(\beta_1 u_1 + \cdots + \beta_r u_r) \quad \text{and} \quad w_0 = \frac{n}{q} u_{r+1} \, .$$

Therefore

$$|w - w_0| \leqslant S C_0^{-\tau(B, U)} \leqslant C_0^{-\tau(B, U)/2}$$

if C_0 is sufficiently large. Hence

$$\begin{aligned} |e(w)^{\lambda_{r+1}} - e(w_0)^{\lambda_{r+1}}| &\leqslant C_5^{SUL} C_0^{-\tau(B, U)/2} \\ &\leqslant C_0^{-\tau(B, U)/3} \end{aligned}$$

provided that

$$SUL \prec (\log B) U^{\kappa} \, .$$

The parts of the monomials involving $e\left(\frac{n}{q} u_j\right)^{\lambda_j}$ for $j = 1, \ldots, r$ is also estimated with an exponent of SUL. This all shows that the measure function $\tau(B, U)$ dominates all the other functions occurring in exponents, and proves the lemma.

Observe that the condition of the lemma provides us with a sufficient lower bound for κ, namely

$$b\sigma + 1 + \frac{r+1}{r+2}\sigma < \kappa \, .$$

In particular,

$$(b + 1)\sigma + 1 \leqslant \kappa$$

is also sufficient. We assume that κ satisfies this condition.

The power b will be determined to satisfy the last step of the proof in the last section of this chapter.

Lemma 2. *Let $N \leqslant S$. Suppose also that $n \leqslant S$ and $|m| \leqslant M$. Let q be a positive integer. Then*

$$D^{(m)} F\left(\frac{n}{q}, \frac{n}{q} u_1, \ldots, \frac{n}{q} u_r; \frac{n}{q} u_{r+1}\right)$$

is an algebraic number of size

$$\leqslant C^{MN+SUL}$$

with an appropriate constant C.

Proof. This is the same trivial estimate as at the beginning of the proof. The only additional feature involved concerns the denominator q. But we note that if α is an algebraic number $\neq 0$, and d is a denominator for α, then d is also a denominator for $\alpha^{1/q}$ (easy verification). Thus the size of $\alpha^{1/q}$ is bounded by the size for α. Of course, the degree of $\alpha^{1/q}$ goes up, and therefore so does the relative height with respect to a number field containing $\alpha^{1/q}$ if $q > 1$, and this degree will have to be dealt with in due course.

§ 5. Extrapolation on Integral Multiples

Lemma. *Let $b \geqslant 1$ be fixed. There exists $0 < \delta < 1$ having the following property. Let v be an integer $\geqslant 0$ and let*

$$S_v = NM^{v\delta}.$$

Suppose $v\delta \leqslant b$. If $|m| \leqslant M/2^v$ and $n \leqslant S_v$, then

$$D^{(m)}F(n, nu_1, \ldots, nu_r; nu_{r+1}) = 0.$$

Proof. By induction on v. For $v = 0$ this is a construction. We assume the result for v. Let

$$|m'| \leqslant M/2^{v+1}$$

and let

$$g(z) = D^{(m')}F(z, zu_1, \ldots, zu_r).$$

Then for $k \leqslant M/2^{v+1}$ we have

$$D^k g(z) = \sum_{|\rho| \leqslant k} b(\rho, u) D^{(m')+(\rho)}F(z, zu_1, \ldots, zu_r)$$

where $b(\rho, u)$ are simple coefficients in which the u_j occur as power products with obvious exponents, easily estimated by U^M.

Since

$$D^{(m)}F(n, nu_1, \ldots, nu_r; nu_{r+1}) = 0$$

for $n \leqslant S_\nu$ and $|m| \leqslant M/2^\nu$, it follows by Lemma 1 of § 4 that

$$|D^{(m)}F(n, nu_1, \ldots, nu_r)| \leqslant C_0^{-\tau(B,U)/4}.$$

Let $n' \leqslant S_{\nu+1}$. We now select a circle of radius

$$R = NM^{(\nu+1)\delta}M^\varepsilon = S_{\nu+1}M^\varepsilon$$

and pick ε, δ such that

$$\varepsilon + \delta + \frac{r+1}{r+2} + \frac{1}{\sigma} < 1 .$$

(This is the main point at which it is significant that L bears a relation to M with exponent < 1.) By the Hermite interpolation formula, applied to the set of positive integers $\leqslant S_\nu$, we get an estimate

$$|g(n')| \leqslant \|g\|_{S_{\nu+1}} \leqslant \frac{\|g\|_R C^{MS_\nu}}{(R/S_{\nu+1})^{MS_\nu}} + (CS_{\nu+1}/S_\nu)^{MS_\nu} C_0^{-\tau(B,U)/4} .$$

Since $S_{\nu+1}/S_\nu = M^\delta$, it follows that the second term on the right is small. On the other hand, easy estimates show that

$$\|g\|_R \leqslant C^{RUL} \quad \text{and} \quad RUL = S_{\nu+1}M^\varepsilon LU .$$

Hence the denominator of the first term wins over the numerator provided that

$$S_{\nu+1}M^\varepsilon LU < MS_\nu ,$$

which amounts precisely to the condition on ε, δ stated above. Thus finally we get the estimate

$$|D^{(m')}F(n', n'u_1, \ldots, n'u_r)| = |g(n')| \leqslant C_7^{-MS_\nu \log M} .$$

By Lemma 1 of the preceding section, a similar estimate applies to the algebraic number obtained in the substituted derivative. On the other hand, we have an estimate for the height of this algebraic number in Lemma 2, and consequently we obtain the upper and lower bounds (in case this algebraic number is $\neq 0$):

$$\frac{1}{C_8^{MN+LS_{\nu+1}U}} \leqslant |D^{(m')}F(n', n'u_1, \ldots, n'u_r; n'u_{r+1})| \leqslant C_9^{-MS_\nu \log M} .$$

But we have both $LS_{\nu+1}U < MS_\nu$, and also MN has lower order of magnitude than $MS_\nu \log M$ by a factor of $\log M$ (a more delicate comparison). This contradiction concludes the inductive step, and also concludes the proof of the lemma.

§ 6. Extrapolation on Fractional Multiples

Lemma. *If $n \leqslant NM$ and q is a prime $\leqslant M$ then*

$$F\left(\frac{n}{q}, \frac{n}{q} u_1, \ldots, \frac{n}{q} u_r ; \frac{n}{q} u_{r+1}\right) = 0 .$$

In other words, the main lemma is true.

Proof. We let S be the largest value attainable in the lemma of § 5. In other words,

$$S = NM^b$$

We may then apply the Hermite interpolation formula, and we get upper bounds as before for $g(n/q)$, where

$$g(z) = F(z, zu_1, \ldots, zu_r) .$$

The algebraic number

$$\xi = F\left(\frac{n}{q}, \frac{n}{q} u_1, \ldots, \frac{n}{q} u_r ; \frac{n}{q} u_{r+1}\right)$$

has a size estimated just as before, but this time, it has degree

$$\ll q^{r+1} \leqslant M^{r+1} .$$

Consequently the final inequalities if $\xi \neq 0$ read

$$\frac{1}{C_{10}^{NMLUq^{r+1}}} \leqslant |\xi| \leqslant C_{11}^{-MS} .$$

This is a contradiction provided that

$$NMLUM^{r+1} < MS = MNM^b ,$$

which tells us what to take for b. For instance, $b = r + 2$ will do.

If we took $\sigma = r + 3$ at the beginning, then we find that the condition

$$(b + 1)\sigma + 1 \leqslant \kappa$$

yields the value

$$\kappa = (r + 3)^2 + 1 .$$

This concludes the proof.

Chapter IX. Linear Combinations of Elliptic Logarithms

In this chapter we give the analogue for elliptic logarithms of a Baker inequality, proved for elliptic curves with complex multiplication. The first inequality of this type was given by Masser [Mas 1]. We shall follow Coates–Lang [CL], giving a stronger one, with a pattern of proof which follows the original Baker arguments more closely, especially in the use of the Kummer theory.

On elliptic curves, Coates [Co 6] was the first to use such arguments, requiring a lower bound for the degree of the field of torsion points of prime order for large primes, or for division points.

The appropriate Kummer statement is now of course known as Bashmakov's theorem (for elliptic curves), and was also later proved by Ribet [Ri] for abelian varieties with complex multiplication.

As pointed out in [CL], we have kept explicit the dependence of constants on the heights of the algebraic points. The theorem obviously extends to abelian varieties with complex multiplication.

There are still some difficulties associated with giving a proof for the stronger result corresponding to the Baker–Feldman let alone the Baker–Tijdeman theorem, let alone for even stronger results as conjectured in [L 5].

For higher dimensional generalizations to abelian varieties with complex multiplication, see [CL], following the work of Masser [Ma 1] and Lang [L 11]. The proof is essentially the same, except that for the final contradiction, it is necessary to combine an important lemma of Masser on polynomials, and an argument of Galois theory, to expand the set of points where the basic polynomial vanishes.

The hypothesis of complex multiplication is made for the proof of the main lemma. At this time no one knows how to make the extrapolation procedure succeed, because the Weierstrass (abelian) functions have a growth of order 2, and one does not see how to construct enough zeros of the approximating auxiliary function to match this without the complex multiplication. Otherwise, the Bashmakov theorem (the rest of the proof) applies to the general case.

§ 1. Remarks on Complex Multiplication

Let A be an elliptic curve with complex multiplication, defined over a number field K. We assume that the imaginary quadratic field k of complex multiplications is contained in K, and we let

$$\mathfrak{o} = \text{End}(\mathbf{C}/\Lambda)$$

where Λ is the period lattice. Each element of $\mathfrak{o}$ corresponds to multiplication by a complex number, which is identified with that element.

We have the exponential map

$$\exp: \mathbf{C}/\Lambda \to A_{\mathbf{C}}$$

given by the Weierstrass functions. For convenience, it is better to deal with functions which are holomorphic at the origin. Thus instead of the Weierstrass functions, we may deal with any other elliptic function f, analytic with non-vanishing derivative at 0, say giving an analytic isomorphism on a disc $\mathbf{D}_{3\rho}(0)$ of radius 3ρ centered at 0. One could take

$$f = \wp/\wp' ,$$

or a suitable translation of the $\wp$-function, so that f, f' still satisfy the old differential equation

$$f'^2 = f^3 + af + b .$$

One can then apply the estimates of Chapter VII, § 3 to this situation. For definiteness and convenience, *we take the latter convention.*

A point $u \in \mathbf{C}$ is called an **algebraic point for the exponential map** if $\exp u$ is algebraic, i.e. if $f(u)$ (and hence $f'(u)$) is algebraic.

The ring $\mathfrak{o}$ has a basis over $\mathbf{Z}$, and we write

$$\mathfrak{o} = [1, \eta_0] = \mathbf{Z} + \mathbf{Z}\eta_0 .$$

For elements of k, the two conjugates have the same absolute value, and thus

$$\|\gamma\| = |\gamma| .$$

For any positive integer S, we denote by $\mathfrak{o}(S)$ the subset of elements $\gamma \in \mathfrak{o}$ such that $\|\gamma\| \leqq S$. If instead of using the above norm we used the sup norm with respect to coordinates relative to $1, \eta_0$, then we could take for such elements all combinations

$$n + n'\eta_0$$

with integers n, n' satisfying $|n|, |n'| \leqq S$. As it is, we see nevertheless that

$$\# \mathfrak{o}(S) \gg\ll S^2 ,$$

where the implicit constant depends only on $\mathfrak{o}$.

Let $u_1, \ldots, u_n \in C$. *Then*

$$\#\{\gamma \in \mathfrak{o}(S),\ \gamma u_j \in \mathbf{D}_\rho(0) \bmod \Lambda \text{ for all } j\} \gg S^2 \rho^{2n},$$

where the implied constant depends only on Λ, n.

Proof. Decompose $\mathbf{C}/\Lambda$ into parallelotopes of width approximately equal to ρ. The number of such parallelotopes is $\gg\ll 1/\rho^2$. There exists a subset Γ_1 of $\mathfrak{o}(S)$ such that

$$\#(\Gamma_1) \gg S^2 \rho^2,$$

and such that all elements γu_1 with $\gamma \in \Gamma_1$ lie in the same small parallelotope. There exists a subset Γ_2 of Γ_1 such that

$$\#(\Gamma_2) \gg S^2 (\rho^2)^2$$

and such that all elements γu_2 with $\gamma \in \Gamma_2$ lie in the same small parallelotope. Inductively, we obtain a subset Γ_n of $\mathfrak{o}(S)$ such that γu_j lie in the same small parallelotope for all $\gamma \in \Gamma_n$ and all j. Pick $\gamma_0 \in \Gamma_n$. Then $(\gamma - \gamma_0) u_j$ lie in a small parallelotope with a corner at 0, thus proving our assertion. (The method was first used in [L 3], Chapter II.)

We also want to deal with conjugates. Each $\gamma \in \mathfrak{o}$ has a corresponding algebraic endomorphism of A, which we denote by γ_A. If (x, y) is a point on A, then

$$\gamma_A(x, y) = (\Phi_\gamma(x), \Omega_\gamma(x, y))$$

where Φ, Ω are rational functions, defined over K. Let

$$\sigma : K \to K^\sigma$$

be an isomorphism of K. Then Φ^σ, Ω^σ are defined by applying σ to the coefficients of these rational functions, and define an endomorphism γ_A^σ of the curve A^σ, obtained from the equation

$$Y^2 = X^3 + a^\sigma X + b^\sigma .$$

For any point $P \in A_K$ we then have

$$(\gamma_A P)^\sigma = \gamma_A^\sigma P^\sigma ,$$

where P^σ is the point obtained by applying σ to the coordinates of P.

Suppose that

$$P_j = \exp u_j, \quad j = 1, \ldots, n$$

and $P_j \in A_K$ for all j, so u_j are algebraic points of the exponential map. We let

$$\mathfrak{o}_\rho(u, S)$$

consist of those elements $\gamma \in \mathfrak{o}(S)$ such that $(\gamma_A P_j)^\sigma$ lies in a disc of radius ρ centered at the origin in $(A^\sigma)_{\mathbf{C}}$ for all conjugates σ of K and all j. Taking ρ small enough, we can assume that the function f^σ also gives an analytic isomorphism on a disc of radius 3ρ centered at the origin.

By repeating the inductive procedure, we then get:

Lemma 1.1. $\# \mathfrak{o}_\rho(u, S) \gg S^2$, *where the constant implicit in* $\gg$ *depends on* ρ, n, A *but not on* S *or* $u_1, \ldots, u_n$. *Furthermore, the values*

$$f(\gamma u_j)^\sigma, \quad f'(\gamma u_j)^\sigma$$

are bounded for all σ, j *and* $\gamma \in \mathfrak{o}_\rho(u, S)$.

We return to the composition of f with elements of $\mathfrak{o}$.

Lemma 1.2. *Let* $\gamma \in \mathfrak{o}(S)$. *There exist polynomials* Φ_γ, Ψ_γ, Ω_γ *of degrees*

$$\ll S^2,$$

with coefficients in K *of heights*

$$\leqslant C_1^{S2}$$

such that

$$f \circ \gamma = \frac{\Phi_\gamma(f)}{\Psi_\gamma(f)} \quad \text{and} \quad f' \circ \gamma = \frac{\Omega_\gamma(f, f')}{\Psi_\gamma(f)}$$

Proof. We write γ in the form

$$\gamma = n + m\eta_0, \quad \text{with} \quad |n|, |m| \ll S\,.$$

We apply Theorems 2.4 and 3.1 of Chapter II. The assertion is then obvious, because multiplication of a point by η_0 and addition are given by fixed rational functions, composed with those arising from the two theorems mentioned above. Note that we can always put $f \circ \gamma$ and $f' \circ \gamma$ over a common denominator $\Psi_\gamma(f)$.

Furthermore, $f' \circ \gamma$ satisfies the Weierstrass equation with respect to $f \circ \gamma$, and is therefore integral over $K[f \circ \gamma]$. If we put Φ_γ/Ψ_γ in lowest form, as rational function in $K(f)$, then its denominator is also a denominator for $f' \circ \gamma$. Doing this does not affect the size of the polynomials and rational functions involved, within the order of magnitude prescribed in the lemma.

In particular, if u *is a point such that* γu *lies in a small neighborhood of* 0 *modulo the period lattice, then* f *is defined at* γu, *and consequently* $\Psi_\gamma(f(u))$ *is* $\neq 0$.

§ 2. Statement of the Theorem

Let $P_1, \ldots, P_n \in A_K$ be algebraic points on A. We let

$$u_j = \log P_j$$

where u_j is selected in a fixed fundamental domain for the period lattice Λ. We let

$$U_j = \log H_K(P_j) \quad \text{and} \quad U = \max U_j .$$

We let

$$\tau(B, U) = (\log B)^{\kappa} U^{\kappa'}$$

where

$$4(r+3)+1 < \kappa \quad \text{and} \quad (r+5)(r+3)+1 < \kappa' ,$$

and $r+1$ is the maximum number of linearly independent elements among $u_1, \ldots, u_n, \omega_1$, over $\mathfrak{o}$. Observe that $\mathfrak{o}\omega_1$ is a lattice, of finite index in Λ.

Theorem 2.1. *Let $\beta_1, \ldots, \beta_n \in K$ and assume*

$$\beta_1 u_1 + \cdots + \beta_n u_n \neq 0 .$$

Let $B = H_K(1, \beta_1, \ldots, \beta_n)$. Then there exists an inequality

$$|\beta_1 u_1 + \cdots + \beta_n u_n| \geqq C_0^{-\tau(B, U)} ,$$

where C_0 is a sufficiently large constant, depending only on n, A, $[K: \mathbf{Q}]$.

The rest of the chapter will provide the proof.

We shall first perform a reduction which allows us to assume that the u_j are linearly independent over $\mathfrak{o}$. We assume that Theorem 2.1 is false, and thus that we have the inequality

$$|\beta_1 u_1 + \cdots + \beta_n u_n| \leqq C_0^{-\tau(B, U)}$$

where C_0 is a sufficiently large constant.

First, we observe that without loss of generality, we may, and shall always assume that u_1 is a period.

Second, we can argue exactly as with ordinary logarithms, in Chapter VIII, § 1, using Theorem 5.2 of Chapter IV, which gives us the possibility of expressing u_j linearly in terms of a maximal set of linearly independent ones. Linear independence is here over $\mathfrak{o}$. Thus inductively, we form the sequence

$$u_1, u_{j2}, \ldots, u_{jr+1}$$

as before. We express the remaining u_j as linear combinations of these elements, and

$$\eta_0 u_1, \eta_0 u_{j2}, \ldots, \eta_0 u_{jr+1}$$

with rational coefficients satisfying the bound $\ll U^r \log\log U$. We can then argue in exactly the same way as before.

We are therefore reduced to proving the following theorem.

Theorem 2.2. *Let* $u_j = \log P_j$ $(j = 1, \ldots, r+1)$ *and let*

$$U = \max U_j .$$

Assume that $u_1, \ldots, u_{r+1}$ *are linearly independent over* $\mathfrak{o}$. *Let* $\beta_1, \ldots, \beta_r \in K$ *and let* $B = H_K(1, \beta_1, \ldots, \beta_r)$. *Then* $u_1, \ldots, u_{r+1}$ *are linearly independent over the algebraic numbers, and we have*

$$|\beta_1 u_1 + \cdots + \beta_r u_r - u_{r+1}| \geqq C_0^{-\tau(B,U)}$$

where C_0 *is a constant depending only on* r, A, $[K:\mathbf{Q}]$.

We make one final reduction. It is sometimes convenient that the points u_j should lie in a small neighborhood of the origin, where the function f gives an analytic isomorphism. We have seen in § 1 that we can find an integer γ such that

$$\gamma u_j \in \mathbf{D}_\rho(0) \bmod \Lambda \quad \text{for } j = 1, \ldots, r+1 ,$$

and γ is bounded in absolute value by a constant depending only on r, ρ, and the period lattice. Replacing u_j by γu_j, that is multiplying the term on the left by γ, and changing the coefficient β_1 of the period u_1 by a bounded term in $\mathfrak{o}$, we see that the proof of the inequality is reduced to that when the points u_j lie in the given neighborhood of 0 modulo the periods. This trick eliminates the dependence of constants on $u_1, \ldots, u_{r+1}$ in various situations, e.g. the transference theorem of Masser [Ma 1] extended by Lang to abelian varieties [L 11], Theorem 6.

§ 3. Main Lemma and its Application

Main Lemma. *Assume that we have the inequality*

$$|\beta_1 u_1 + \cdots + \beta_r u_r - u_{r+1}| < C_0^{-\tau(B,U)}$$

for a sufficiently large constant C_0. Then there exists a polynomial, not identically zero,

$$P(X_1, \ldots, X_{r+1}) = \sum a_{(\lambda)} X_1^{\lambda_1} \ldots X_{r+1}^{\lambda_{r+1}}$$

satisfying:

(i) *The coefficients $a_{(\lambda)}$ are integers.*
(ii) $0 \leqslant \lambda_1, \ldots, \lambda_{r+1} \leqslant L$, *and L is sufficiently large.*
(iii) *We have*

$$P\left(f\left(\frac{1}{q} u_1\right), \ldots, f\left(\frac{1}{q} u_{r+1}\right)\right) = 0$$

for some prime $q > L$, and also such that q remains prime in the field of complex multiplication.

To apply the main lemma, we let

$$K_s = K\left(f\left(\frac{1}{q} u_1\right), \ldots, f\left(\frac{1}{q} u_s\right)\right).$$

We have by Bashmakov's theorem,

$$[K_s : K_{s-1}] \gg q^2 .$$

The same lemma as in Chapter VIII, § 2 shows that P is identically zero, a contradiction which proves the theorem.

The main lemma will be proved in the subsequent sections.

Note that instead of requiring that q remains prime in the field of complex multiplication, we needed only to pick q so large that Bashmakov's theorem would yield

$$[K_s : K_{s-1}] > L .$$

In the case of complex multiplication, it could happen that the order of magnitude of the degree $[K_s : K_{s-1}]$ is q rather than q^2.

§ 4. Construction of the Approximating Function

Let $a_{(\lambda)}$ denote integers to be determined, and let

$$F(z_1, \ldots, z_r) = \sum a_{(\lambda)} f(z_1)^{\lambda_1} \ldots f(z_r)^{\lambda_r} f(\beta_1 z_1 + \cdots + \beta_r z_r)^{\lambda_{r+1}} .$$

where $0 \leqslant \lambda_1, \ldots, \lambda_{r+1} \leqslant L$. We let

$$N = [(\log B)U] \quad \text{and} \quad M = N^a U^t$$

where a, t are treated as unknowns, determined as we go along to make the method of proof successful. (It turns out $a = 4$, $t > r + 1$ will do.)

The use of a factor U in the definition of N is to take care of the last step in the proof in § 6. We could introduce U additively for the same purpose, i.e. put

$$N = [\log B + U],$$

but to get substantially better results in line with Baker–Feldman–Tijdeman, the whole structure has to be refined using other interpolating polynomials, so no attempt to get best possible results out of the present structure has been made.

We note that F is equal to a polynomial $P(X_1, \ldots, X_{r+1})$ evaluated at

$$X_j = f(z_j) \quad \text{for } j = 1, \ldots, r$$
$$X_{r+1} = f(\beta_1 z_1 + \cdots + \beta_r z_r).$$

Let

$$D^{(m)} = D_1^{m_1} \ldots D_r^{m_r}$$

be a differential operator. Then

$$D^{(m)} F = P_{(m)}(\mathbf{f}(z_1), \ldots, \mathbf{f}(z_r), \mathbf{f}(\beta_1 z_1 + \cdots + \beta_r z_r))$$

where

$$\mathbf{f} = (f, f').$$

If z_{r+1} is a new variable, we use the notation

$$D^{(m)} F(z_1, \ldots, z_r; z_{r+1}) = P_{(m)}(\mathbf{f}(z_1), \ldots, \mathbf{f}(z_r), \mathbf{f}(z_{r+1})).$$

This is the expression obtained by substituting z_{r+1} for $\beta_1 z_1 + \cdots + \beta_r z_r$. In particular, if $\eta \in k$ and ηu_j is not a pole of f, then

$$D^{(m)} F(\eta u \ \ldots, \eta u_r; \eta u_{r+1})$$

is an algebraic number. If in addition $\eta \in \mathfrak{o}$ then this algebraic number lies in K.

We require that

$$D^{(m)} F(\eta u_1, \ldots, \eta u_r; \eta u_{r+1}) = 0$$

for

$$\eta \in \mathfrak{o}_p(u, N) \quad \text{and} \quad |m| \leqslant M .$$

This amounts to a system of linear equations for the $a_{(\lambda)}$, in the field

$$K_0(\beta_1, \ldots, \beta_r, P_1, \ldots, P_{r+1})$$

where K_0 is a field of definition for A, also containing k. This field has a basis over K_0 consisting of power products of β_j, and coordinates for P_j. These have size

$$\ll B^c\, e^{cU}$$

for some constant c. In these linear equations, we have:

Number of variables $= L^{r+1}$

Number of equations $\ll M^r N^2$

Size of coefficients $\leqslant M^M C_1^{M+L} C_1^{N^2(M+L)U} B^M$.

This last estimate is immediate from Chapter VII, § 3. The power B^M arises from the chain rule with respect to the $\beta_j z_j$ $(j = 1, \ldots, r)$, and from our definition of N is like C^{MN} for some constant C.

In the Siegel Lemma 2 of Chapter VII, § 2 we wish to make the number of variables so large as to end up with coefficients which have size $\leqslant C^{MN}$ with a suitable constant C. For this, we therefore make

$$L^{r+1} = M^r N^2 NU \cdot [K : \mathbf{Q}] C_2 \tag{1}$$

where C_2 is an appropriately large constant. Then the Dirichlet exponent in that Lemma 2 turns out to be $\ll 1/NU$. This achieves our purpose. It also determines L in terms of M, and U, i.e.

$$\boxed{L \gg\ll M^{r/(r+1)} N^{3/(r+1)} U^{1/(r+1)} .}$$

For convenience of notation, if we deal with quantities X, Y such that

$$X = N^b U^t X' \quad \text{and} \quad Y = N^{b'} U^{t'} Y'$$

where X', Y' do not contain N, U, then we write $X \prec Y$ to mean $b < b'$ and $t + b < t' + b'$.

With this notation, we require that

$$\boxed{LU < M.} \tag{2}$$

This amounts to $a > 3$ and

$$\frac{3}{r+1} + \frac{1}{r+1} + 1 < (a+t)\frac{1}{r+1}.$$

For definiteness we select

$$\boxed{a = 4,} \tag{3}$$

in which case this leaves us with the condition

$$\boxed{t > r+1.} \tag{4}$$

We shall need no further condition on t.

§ 5. Some Estimates

In the first estimate, we want to see that a polynomial in values

$$f\left(\frac{\gamma}{q}u_j\right),\ f\left(\frac{\gamma}{q}(\beta_1 u_1 + \cdots + \beta_r u_r)\right)$$

does not differ much when we substitute

$$f\left(\frac{\gamma}{q}u_{r+1}\right) \quad \text{for } f\left(\frac{\gamma}{q}(\beta_1 u_1 + \cdots + \beta_r u_r)\right).$$

Lemma 5.1. *Let $\gamma \in \mathfrak{o}(S)$. Let q be a positive integer. Assume that $\frac{\gamma}{q}u_j$ lies in the ball of radius ρ, modulo the period lattice for each $j = 1, \ldots, r+1$. Also assume that*

$$MN \leqslant \tau(B, U), \qquad \log S \leqslant \tau(B, U).$$

Then for $|m| \leqslant M$ we have for C_0 sufficiently large:

$$\left|D^{(m)}F\left(\frac{\gamma}{q}u_1, \ldots, \frac{\gamma}{q}u_r; \frac{\gamma}{q}u_{r+1}\right) - D^{(m)}F\left(\frac{\gamma}{q}u_1, \ldots, \frac{\gamma}{q}u_r\right)\right| \leqslant C_0^{-\tau(B,U)/4}.$$

Proof. Write

$$w_{r+1} = \frac{\gamma}{q} u_{r+1} \quad \text{and} \quad w = \frac{\gamma}{q}(\beta_1 u_1 + \cdots + \beta_r u_r) .$$

Then

$$|w_{r+1} - w| \leqslant S C_0^{-\tau(B,U)/2} .$$

Since $\mathbf{f}$ satisfies a Lipschitz condition on the ball of radius 3ρ, we obtain

$$|f(w_{r+1}) - f(w)| \leqslant C_0^{-\tau(B,U)/3} ,$$

and similarly for f' instead of f.

The difference of two monomials involving $f(w)$ and $f(w_{r+1})$ is easily estimated. Indeed, for a positive integer λ we have

$$X^\lambda - Y^\lambda = (X - Y)(X^{\lambda-1} + \cdots + Y^{\lambda-1}) .$$

We put $X = f(w)$ and $Y = f(w_{r+1})$, and similarly with f' instead of f. By assumption, these values are bounded independently of γ, q because w, w_{r+1} lie in a small neighborhood of the origin. Consequently we obtain from the lemma of Chapter VII, § 3:

$$\begin{aligned}\left| P_{(m)}\left(\mathbf{f}\left(\frac{\gamma}{q}u\right); f(w_{r+1})\right) - P_{(m)}\left(\mathbf{f}\left(\frac{\gamma}{q}u\right); \mathbf{f}(w)\right)\right| &\leqslant C_3^{MN} B^M C_0^{-\tau(B,U)/3} \\ &\leqslant C_0^{-\tau(B,U)/4} ,\end{aligned}$$

as was to be shown.

Remark. The parameter U played no role in the previous lemma, involving only complex estimates, and no heights. The next estimate does involve heights.

We write

$$\mathbf{f} = (1, f_1, f_2)$$

for convenience, so $f = f_1$ and $f' = f_2$. Let w be a point such that $\mathbf{f}(qw)$ is rational over K for some positive integer q. Then

$$[K(\mathbf{f}(w)) : \mathbf{Q}] \leqslant q^2 [K : \mathbf{Q}].$$

We shall use the quadratic form of Néron–Tate to get estimates for the heights of points $\mathbf{f}\left(\frac{\gamma}{q} u\right)$ where u is an algebraic point of the exponential map. Let $h = h_\wp$ be the absolute height, and $\hat{h}$ the quadratic form such that

$$h = \hat{h} + O(1). \tag{1}$$

We write simply $P(z) = \exp z = (1, \wp(z), \wp'(z))$. Since we took f to be a translation of $\wp$ by some fixed point of finite order, then we have the estimate for any algebraic point z of the exponential map (i.e. such that $P(z)$ is algebraic):

$$h(f(z)) \gg\ll h(P(z)) + O(1). \tag{2}$$

Let $1, \eta_0$ be a basis of $\mathfrak{o}$ over $\mathbf{Z}$. For $\gamma \in \mathfrak{o}$ write

$$\gamma = s + s'\eta_0 .$$

If $\gamma \in \mathfrak{o}(S)$ then $|s|, |s'| \ll S$. From the bilinearity, we get

$$\hat{h}\left(P\left(\frac{s}{q} z\right)\right) = \frac{s^2}{q^2} \hat{h}(P(z)). \tag{3}$$

Since $h(P(\eta_0 z)) \ll h(P(z)) + O(1)$ and

$$P\left(\frac{\gamma}{q} u\right) = P\left(\frac{s}{q} u\right) + P\left(\frac{s'}{q} \eta_0 u\right)$$

we get the estimate

$$h\left(P\left(\frac{\gamma}{q} u\right)\right) \ll \frac{s^2}{q^2} (h(P(u)) + O(1)). \tag{4}$$

The terms $O(1)$ depend only on the elliptic curve A.

Let $E = K\left(\mathbf{f}\left(\frac{u}{q}\right)\right)$ so that $[E:\mathbf{Q}] \ll q^2$. Since

$$h_E = [E:\mathbf{Q}]h,$$

we may multiply the estimate of (4) by q^2, exponentiate, and compare the size to the relative height to obtain an estimate on the size, namely:

$$\boxed{\operatorname{size} f\left(\frac{\gamma}{q} u\right) \leqslant C_4^{S^2 U + q^2}.}$$

where

$$U = \log H_K(f(u)).$$

Lemma 5.2. *Let $\gamma \in \mathfrak{o}(S)$, and let q be a positive integer such that $\frac{\gamma}{q} u_j$ is not a pole of f for all j. Then the algebraic number*

$$D^{(m)}F\left(\frac{\gamma}{q} u_1, \ldots, \frac{\gamma}{q} u_r; \frac{\gamma}{q} u_{r+1}\right)$$

is either equal to 0, or its size is bounded by

$$C_5^{M(S^2 U + q^2)}.$$

This estimate is a "Liouville" type estimate. In the inductive procedure to be followed, because of additional conditions, one can replace a term $S^2 M$ in the exponent by $S^2 L$. This is crucial for the induction, and the result is stated precisely in the next lemma.

Lemma 5.3 (Baker–Coates Lemma). *Let (m) be such that $|m| \leqq M$. Let $\gamma \in \mathfrak{o}_\rho(u, S)$. Assume that*

$$D^{(\mu)}F(\gamma u_1, \ldots, \gamma u_r; \gamma u_{r+1}) = 0$$

for all (μ) with $|\mu| < |m|$. Then either

$$D^{(m)}F(\gamma u_1, \ldots, \gamma u_r; \gamma u_{r+1}) = 0,$$

or:

(i) *Its conjugates are bounded by*

$$M^M C^{M+L} B^M C^{MN} \leqq C_6^{MN}.$$

(ii) *Its denominator is bounded by*

$$M^M C^{M+L} B^M C^{(S^2 L + M)U} S^{cM} \leqq C_7^{MN} C_8^{(S^2 L + M)U} S^{cM}$$

Proof. We know that $D^{(m)}F$ is a polynomial $P_{(m)}$ as usual, and we have an estimate for the degree and coefficients of this polynomial. Under our special assumption on γ, we know that the values $f_i(\gamma u_j)$ $(j = 1, \ldots, r + 1)$ are bounded, so that when we substitute these values in the polynomial, we get a number for which estimate (i) holds.

The problem is to deal with denominators. Let

$$\begin{aligned} E(z_1, \ldots, z_r) &= F(\gamma z_1, \ldots, \gamma z_r) \\ &= \sum a_{(\lambda)} \prod f(\gamma z_j)^{\lambda_j} f(\gamma \beta_1 z_1 + \cdots + \gamma \beta_r z_r)^{\lambda_{r+1}}. \end{aligned}$$

Lemma 2 of § 1 gives us an expression for $f(\gamma z_j)$ as rational function in $f(z_j)$. It allows us to construct a generic denominator. Let

$$G(z_1, \ldots, z_r) = \left(\prod_{j=1}^{r} \Psi_\gamma(f(z_j))^L \right) \Psi_\gamma(f(\beta_1 z_1 + \cdots + \beta_r z_r))^L .$$

Then GE is a polynomial in the functions

$$f(z_j); \quad f(\beta_1 z_1 + \cdots + \beta_r z_r) .$$

Its degree is $\ll S^2 L$, and its coefficients have size $\ll C^{S^2 L}$.

The derivative $D^{(m)}(GE)(z_1, \ldots, z_r)$ can be expressed as a sum of terms $D^{(\mu)} G D^{(m-\mu)} E$, with binomial coefficients. Furthermore,

$$D^{(\mu)} E(u_1, \ldots, u_r; u_{r+1}) = \gamma^{(\mu)} D^{(\mu)} F(\gamma u_1, \ldots, \gamma u_r; \gamma u_{r+1}) ,$$

where $\gamma^{(\mu)}$ is a monomial of order $|\mu|$. The latter expression vanishes by assumption if $|\mu| < |m|$. Consequently

$$\begin{aligned} D^{(m)}(GE)(u_1, \ldots, u_r; u_{r+1}) &= G(u_1, \ldots, u_r; u_{r+1}) D^{(m)} E(u_1, \ldots, u_r; u_{r+1}) \\ &= G(u_1, \ldots, u_r; u_{r+1}) \gamma^{(m)} D^{(m)} F(\gamma u_1, \ldots, \gamma u_r; \gamma u_{r+1}) . \end{aligned}$$

Therefore the denominator which we want to estimate for (ii) is a product of:

(a) a denominator for $D^{(m)}(GE)(u_1, \ldots, u_r; u_{r+1})$
(b) $G(u_1, \ldots, u_r; u_{r+1})$
(c) $\gamma^{(m)}$.

By the lemma of Chapter VII, § 3, we know that

$$D^{(m)}(GE)(u_1, \ldots, u_r; u_{r+1})$$

is a polynomial in the numbers

$$f_i(u_j) , \quad i = 1, 2; \quad j = 1, \ldots, r + 1 ,$$

and this polynomial satisfies:

$$\text{degree} \ll S^2 L + M$$

$$\text{size of coefficients} \leqslant C^{S^2 L} M^M C^M B^M C^{MN} .$$

As the values $f_i(u_j)$ have height $\leqslant C^U$ we see that term (a) introduces a denominator with the required bound $C^{(S^2 L + M) U} C^{MN}$.

Using the fact that $H(\xi) = H(\xi^{-1})$ for any algebraic number ξ, we see that a similar bound prevails for term (b). Finally, we note that

$$H(\gamma^{(m)}) \leqslant H(\gamma)^M, \quad \text{and} \quad H(\gamma) \ll S^{\deg \gamma}.$$

Thus term (c) gives a contribution within the desired bound. This proves the lemma.

Remark. The estimates of this section have nothing to do with complex multiplication and are obviously valid for any elliptic curve.

§ 6. Extrapolation on Integral Multiples

Lemma 6.1. *For each positive integer v, let $S_v = NM^{v\delta}$. We can choose δ sufficiently small such that*

$$D^{(m)}F(\gamma u_1, \ldots, \gamma u_r; \gamma u_{r+1}) = 0$$

for all $\gamma \in \mathfrak{o}_\rho(u, S_v)$ and $|m| \leqslant M/2^v$, for values of v such that

$$v\delta \leqslant r + 2 .$$

Proof. For $v = 0$, this is a construction. We assume that statement for v and prove it for $v + 1$. Suppose the assertion is false for some $\gamma' \in \mathfrak{o}_\rho(u, S_{v+1})$. Let (m') be such that $|m'|$ is the smallest value $\leqslant M/2^{v+1}$ such that

$$D^{(\mu)}F(\gamma' u_1, \ldots, \gamma' u_r; \gamma' u_{r+1}) = 0 \quad \text{if } |\mu| < |m'|$$

but

$$D^{(m')}F(\gamma' u_1, \ldots, \gamma' u_r; \gamma' u_{r+1}) \neq 0 .$$

Let θ be an entire function of order 2 such that θf and $\theta f'$ are entire (θ is obtained trivially from the sigma function). Let

$$\Theta(z_1, \ldots, z_r) = [\theta(z_1)\ldots\theta(z_r)\theta(\beta_1 z_1 + \cdots + \beta_r z_r)]^L .$$

Then ΘF is an entire function. Let

$$g(z) = D^{(m')}(\Theta F)(zu_1, \ldots, zu_r) .$$

Then g is a function of one variable. We have

$$D^n g(z) = \sum_{|\sigma| < n} b(\sigma, u) D^{(m') + (\sigma)}(\Theta F)(zu_1, \ldots, zu_r)$$

where $b(\sigma, u)$ is a simple polynomial in the numbers u and binomial coefficients, easily estimated. We take $n \leqslant M/2^{\nu+1}$. By Lemma 5.1 and a trivial estimate for $D^{(\mu)}\Theta$, obtained from Cauchy's formula for derivatives, we obtain

$$|D^n g(\gamma)| \leqslant C_0^{-\tau(B,U)/4}, \quad \text{all } \gamma \in \mathfrak{o}_\rho(u, S_\nu)\,.$$

We apply the Hermite formula, with radius

$$R = S_{\nu+1} M^{\varepsilon}.$$

Again using Cauchy's formula on F to estimate derivatives of F, we obtain the estimate

$$\|g\|_R \leqslant C^{R^2 LU} C^{MN} \leqslant C_9^{R^2 LU}.$$

This yields the estimate

$$|g(\gamma')| \leqslant \frac{\|g\|_R C^{MS_\nu^2}}{(R/S_{\nu+1})^{MS_\nu^2}} + (CS_{\nu+1})^{MS_\nu^2} \max_{\gamma \in \mathfrak{o}_\rho(u,\, S_\nu)} |D^n g(\gamma)|\,.$$

The second term on the right is small, in fact $\leqslant C_0^{-\tau(B,U)/5}$ for C_0 sufficiently large, provided that for the maximal value S of S_ν we have the dominance

$$(5) \qquad \boxed{MS^2 \prec \tau(B, U)\,,}$$

which is indeed true, it amounts to

$$MNM^{\nu\delta} \prec (\log B)^{\kappa} U^{\kappa'},$$

which in terms of our parameters yields lower bounds for κ, κ':

$$(5') \qquad \boxed{a(1 + \nu\delta) + 1 < \kappa} \quad \text{and} \quad \boxed{(a + t)(1 + \nu\delta) + 1 < \kappa'\,.}$$

We want the denominator of the first term to grow more rapidly than the numerator. Since

$$R/S_{\nu+1} = e^{\varepsilon \log M},$$

the term $C^{MS_\nu^2}$ is overwhelmed by the denominator (through the appearance of

$\log M$). We further need

$$\boxed{R^2LU = S_{\nu+1}^2 M^{2\varepsilon}LU < MS_\nu^2 .}$$

Since $LU < M$ from the beginning, we can clearly pick ε and δ small enough such that

$$(S_{\nu+1}/S_\nu)^2 M^{2\varepsilon}LU < M, \text{ that is } M^{2\delta+2\varepsilon}LU < M ,$$

because $S_{\nu+1}/S_\nu = M^\delta$. Thus finally we obtain the estimate

$$|g(\gamma')| \leqslant \frac{1}{C_{10}^{MS_\nu^2 \log M}}$$

with an appropriate constant C_{10}.

We have obtained an upper bound for $|g(\gamma')|$, showing that it is quite small. From it we obtain an upper bound for $D^{(m')}F$. We can write

$$\begin{aligned} g(\gamma') = {} & \Theta(\gamma' u_1, \ldots, \gamma' u_r) D^{(m')} F(\gamma' u_1, \ldots, \gamma' u_r) \\ & + \sum_{|\mu| < |m'|} c(\mu) D^{(m') - (\mu)} \Theta(\gamma' u_1, \ldots, \gamma' u_r) D^{(\mu)} F(\gamma' u_1, \ldots, \gamma' u_r) . \end{aligned}$$

By the difference Lemma 5.1 again, since

$$D^{(\mu)}F(\gamma' u_1, \ldots, \gamma' u_r, \gamma' u_{r+1}) = 0 ,$$

we conclude that the second sum on the right is bounded by

$$C_0^{-\tau(B,U)/4} .$$

On the other hand, since $\gamma' u_j$ is away from the zeros of θ, it follows immediately from the functional equation of the sigma function (with quadratic growth) that

$$\log |\Theta(\gamma' u_1, \ldots, \gamma' u_r)| \gg -S_{\nu+1}^2 L .$$

Hence we obtain the estimate

$$\log |D^{(m')}F(\gamma' u_1, \ldots, \gamma' u_r)| \ll -MS_\nu^2 .$$

By Lemma 5.1 once more, the substituted derivative satisfies

$$\log |D^{(m')}F(\gamma' u_1, \ldots, \gamma' u_r; \gamma' u_{r+1})| \ll -MS_\nu^2$$

By the Baker–Coates Lemma 5.3, we find the lower inequality

$$-(MN + S_{\nu+1}^2 LU + MU) \ll \log |D^{(m')}F(\gamma' u_1, \ldots, \gamma' u_r; \gamma' u_{r+1})| \ll -MS_\nu^2 .$$

The extreme inequalities are contradictory to each other by the way we have picked our parameters, and the lemma is proved. Observe that we need here, in addition to $LU < M$, that

$$\boxed{MU < MS_v^2\,.} \tag{6}$$

The most dangerous step here is the first one, when $S_0 = N$. Thus if the present pattern of proof is to be followed, some occurrence of U must appear in the original definition of N.

§ 7. Extrapolation on Fractional Multiples

Main Lemma. *Let q be a positive integer $\leqslant M$. Then*

$$F\left(\frac{1}{q}u_1, \ldots, \frac{1}{q}u_r; \frac{1}{q}u_{r+1}\right) = 0\,.$$

Proof. This time, we let

$$g(z) = F(zu_1, \ldots, zu_r)\,.$$

We let S be the maximal value of S_v occurring in the lemma of § 6. The function g has therefore small derivatives of order $\gg M$ at $\gg S^2$ points. From the formula

$$\left|g\left(\frac{1}{q}\right)\right| \leqslant \frac{\|g\|_R C^{MS^2}}{C^{MS^2 \log M}} + (CS)^{MS^2} \max_{\gamma \in \mathfrak{o}_\rho(u,S)} |D^n g(\gamma)|\,,$$

taken with a radius $R = SM^\varepsilon$, we conclude as before that

$$\left|g\left(\frac{1}{q}\right)\right| \leqslant \frac{1}{C^{MS^2 \log M}},$$

and therefore that

$$\log\left|F\left(\frac{1}{q}u_1, \ldots, \frac{1}{q}u_r; \frac{1}{q}u_{r+1}\right)\right| \ll -MS^2\,.$$

We do not have to use Baker–Coates here, but merely the Liouville type estimate of Lemma 5.2 for the size of the algebraic number in that lemma, with $\gamma = 1$. Since this algebraic number has degree $\ll q^{2(r+1)}$, we obtain also the lower inequality

$$-M(q^2 + U)q^{2(r+1)} - MN \ll \log\left|F\left(\frac{1}{q}u_1, \ldots, \frac{1}{q}u_r; \frac{1}{q}u_{r+1}\right)\right| \ll -MS^2\,.$$

We get a contradiction provided that

$$MN + M(q^2 + U)q^{2(r+1)} < MS^2 \tag{7}$$

and certainly if

$$M^{2r+4} < S^2 = (NM^{\nu\delta})^2 \tag{7'}$$

for the maximal value of $\nu\delta$ in the extrapolation of Lemma 6.1. This tells us exactly how far we had to push $\nu\delta$, and it suffices that

$$r + 2 \leqslant \nu\delta . \tag{7''}$$

This concludes the proof.

Note: The way one arrives at the values of the parameters is to go through exactly the steps (1) through (7), solving for the appropriate inequalities which make the method work. Sufficiently high values of κ, κ' are determined by (5).

Introduction to Chapters X and XI

Currently known results concerning lower bounds for linear forms in logarithms are still quite far from best possible results. Some conjectures were given in [L 5], but in that paper, the algebraic numbers α_j are regarded as fixed. If one wants to take them into account, one has the following conjectures of Lang–Waldschmidt. Let $a_1, \ldots, a_n$ be rational numbers > 0, and $b_1, \ldots, b_n$ integers. For $j = 1, \ldots, n$ let

$$B_j = \max\{|b_j|, 1\}, \qquad A_j = \max\{H_{\mathbf{Q}}(a_j), 1\}.$$

$$B = \max B_j. \qquad b_1 \log a_1 + \cdots + b_n \log a_n \neq 0.$$

Conjecture 1. *Let $\varepsilon > 0$. There exists $C(\varepsilon) > 0$ depending only on ε, such that*

$$|b_1 \log a_1 + \cdots + b_n \log a_n| > \frac{C(\varepsilon)^n B}{(B_1 \cdots B_n A_1^2 \cdots A_n^2)^{1+\varepsilon}}.$$

To motivate the conjecture, suppose that $B_1, \ldots, B_n, A_1, \ldots, A_n$ are sufficiently large. Consider the set of numbers

$$b_1 \log a_1 + \cdots + b_n \log a_n$$

with $b_j \in \mathbf{Z}$, $a_j \in \mathbf{Q}$, $a_j > 0$, $|b_j| \leqslant B_j$ and $H_{\mathbf{Q}}(a_j) \leqslant A_j$. This set has cardinality

$$\gg\ll B_1 \cdots B_n A_1^2 \cdots A_n^2$$

where the constant implicit in $\gg\ll$ is absolute. This set is contained in the interval

$$[-nB \log A, nB \log A], \quad \text{with } A = \max A_j.$$

If this set is uniformly distributed in this interval, then the distance from 0 to the closest non-zero element would be

$$\gg\ll nB \log A / B_1 \cdots B_n A_1^2 \cdots A_n^2.$$

On the other hand, the Dirichlet box principle shows that, when $b_1, \ldots, b_n$ are fixed or when $a_1, \ldots, a_n$ are fixed, then there exists a non-zero linear form in the logarithms which is

$$\ll nB \log A/A_1^2 \cdots A_n^2 \text{ in the first case },$$

$$\ll nB \log A/B_1 \cdots B_n \text{ in the second case }.$$

Thus the conjecture is motivated from the uniform distribution and the Dirichlet box principle.

Conjecture 2. *Suppose $a_j \in \mathbf{Z}$, for $j = 1, \ldots, n$. There exists $C(\varepsilon) > 0$ such that*

$$|b_1 \log a_1 + \cdots + b_n \log a_n| > \frac{C(\varepsilon)^n B}{(B_1 \cdots B_n A_1 \cdots A_n)^{1+\varepsilon}}.$$

When the a_j are replaced by elements α_j in a number field K, or by algebraic integers in a number field, one has similar conjectures, since one knows how to count the number of elements of bounded height asymptotically, cf. Schanuel [Sch]. The constant then depends on ε and K. Similarly on an elliptic curve when the α_j are replaced by algebraic points P_j, and $u_j = \log P_j$ are the elliptic logarithms as in [L 5]. Thus if an elliptic curve A is defined over K, let $\| \ \|$ denote the metric on the torus $A_{\mathbf{C}} = \mathbf{C}$ modulo the period lattice. One should have

$$\|b_1 u_1 + \cdots + b_n u_n\| > \frac{C(K, A, \varepsilon)^n}{B^{n+\varepsilon} h^{rn/2+\varepsilon}}$$

where r is the rank of A_K,

$$B = \max |b_j| \quad \text{and} \quad h = \max h_K(P_j).$$

The motivation is carried out similarly, taking into account the obvious estimate for the number of points of height $\leqslant h$ with $h \to \infty$, obtained from the Néron–Tate quadratic form.

From Conjecture 2, we immediately deduce a generalization of Hall's conjecture, namely if $x^p - y^q \neq 0$ with integers $x, y > 0$ then

$$|x^p - y^q| \gg \max (x^p, y^q)^{1 - 1/p - 1/q - \varepsilon}.$$

Indeed, we look at x^p/y^q which is close to 1 if $x^p - y^q$ is small, and take the logarithm, applying Conjecture 2 to get the bound on the right.

At the moment, the theory is very far from conjectures like the above, and we shall briefly summarize where current efforts have led.

Baker [Ba 2] obtained a strengthening of the theorem of Chapter VIII by getting the correct power in the exponent of $\tau(B, U_1, \ldots, U_n)$ with respect to the largest variable, namely the first power. In other words, he proved the inequality

$$|\beta_1 u_1 + \cdots + \beta_n u_n| \geqq C_0^{-\tau(B, U_n)}$$

in the case when $\beta_1, \ldots, \beta_n$ are *rational*, with the function

$$\tau(B, U_n) = (\log B)U_n ,$$

and C_0 depends on $U_1, \ldots, U_{n-1}$. Tijdeman, following the same pattern of proof, took into account explicitly the dependence on $U_1, \ldots, U_{n-1}$, showing that some low power of these numbers could be taken. Thus he obtains an inequality

$$|\beta_1 u_1 + \cdots + \beta_n u_n| \geqslant C_0^{-\tau(B,u)}$$

where

$$\tau(B, u) = (\log B)U_n U_{n-1}^{\theta} ,$$

$U_{n-1} = \max(U_1, \ldots, U_{n-1})$, and θ is a low number, depending on n. The constant C_0 is then sufficiently large, depending only on n and the degree $[K:\mathbf{Q}]$.

At the end of his proof, Tijdeman uses a further improved inequality of Baker [Ba 3], but it is possible to organize the proof so as to avoid this, and rely for what is essentially an induction only on the Baker–Feldman result of Chapter VIII. This inductive technique has interest for its own sake, and for the possibility of application in other contexts, e.g. elliptic curves. We present it in Chapter X.

In [Ba 3], Baker developed a new elimination procedure which enabled him to derive

$$|\beta_1 u_1 + \cdots + \beta_n u_n| \geqslant C_0^{-\tau(B,u)}$$

where

$$\tau(B, u) = (\log B)U_1 \ldots U_n \log U_n .$$

The dependence on U_n to the first power is here lost, but Van der Poorten [VdP 1] noted that $\log U_n$ can be replaced by $\log U_{n-1}$, thus including Tijdeman's inequality.

We shall prove this inequality in Chapter XI, namely with the function

$$\boxed{\tau(B, u) = (\log B)U_1 \ldots U_n \log U_{n-1}}$$

by using the method of Cijsouw–Waldschmidt [C–W], which avoids the extrapolation on integral multiples. The Baker–Feldman theorem gives the correct exponent 1 for $\log B$, but still too high an exponent for U_j. Baker [Ba 2] obtained the correct exponent for U_r, provided $\beta_1, \ldots, \beta_r$ are rational numbers. In the general case (algebraic β_j) the best known result to date [Ba 7], [C–W] is

$$|\beta_0 + \beta_1 u_1 + \cdots + \beta_r u_r| > \exp\{-C(\log B + \log U_r)U_1 \ldots U_r \log U_{r-1}\} .$$

The importance of getting exactly the first power on U_n was shown by Tijdeman in that he was able to settle effectively the classical Catalan conjecture, whether the equation

$$x^p - y^q = 1$$

has only a finite number of solutions in integers p, q, x, y. The difficulty was of course to bound p, q. (For fixed p, q the equation is superelliptic and can be treated by the methods of Siegel–Chabauty–Kubert–Lang, reducing to the elliptic case, for instance.)

We shall now reproduce Tijdeman's arguments to show the reader how the sharp inequality is applied. We assume that p, q are odd primes (the case when one of them is 2 can be treated separately). We have to show p, q are bounded.

We first analyse a divisibility property. We can write

$$x^p = y^q + 1 = (y+1)(y^{q-1} - y^{q-2} + \cdots + 1)\,.$$

The two factors on the right have a g.c.d. equal to 1 or q. Indeed, if l is a prime dividing both of them, so that

$$y \equiv -1 \pmod{l}$$

then

$$y^{q-1} - y^{q-2} + \cdots + 1 \equiv -q \pmod{l}$$

and therefore $l = q$. It then also follows that only the first power of q can divide both terms in the factorization. Arguing similarly on $y^q = x^p - 1$, we conclude: There exist $\delta, \delta' = 0, 1, -1$ such that:

$$x = p^{\delta}X^q + 1 \quad \text{and} \quad y = q^{\delta'}Y^p - 1$$

with some integers X, Y.

There is an obvious symmetry between the terms in x and y. For concreteness, we assume, say, that

$$q \leqslant p \quad \text{and} \quad 0 < x \leqslant y\,.$$

The other cases would be treated the same way.

We carry out the proof in two steps.

Step 1. *$q \ll (\log p)^c$ for some constant c.*

Proof. We note first that

$$(p^{\delta}X^q)^p \quad \text{and} \quad (q^{\delta'}Y^p)^q$$

have the same order of magnitude. We also have:

$$\begin{aligned}&|\log (x-1)^p - \log (y+1)^q|\\ &\qquad \leqslant |\log (x-1)^p - \log x^p| + |\log x^p - \log y^q| + |\log y^q - \log (y+1)^q|\\ &\qquad \ll \frac{p^2}{X^q} = e^{-q\log X + 2\log p}.\end{aligned}$$

Using the factorizations for $x-1$ and $y+1$ obtained above, we find:

$$|p\log(p^\delta X^q) - q\log(q^{\delta'}Y^p)| \ll p^2/X^q$$

whence

$$|p\delta\log p - q\delta'\log q + pq\log(X/Y)| \ll p^2/X^q.$$

By the Baker–Tijdeman inequality, we must have

$$C_0^{-[(\log p)(\log q)]^\theta \log X} \leqslant C_1\, e^{-q\log X + 2\log p}.$$

The exponential term with $2\log p$ on the right can be brought to the left, and having $\log X$ to the first power exactly on both sides allows us to cancel $\log X$. (This is where the sharp inequality is used.) Since $q \leqslant p$ we obtain an inequality

$$q \ll (\log p)^c$$

for some constant c, as desired.

Step 2. $p \ll (\log p)^c$ *for some constant* c.

Proof. This time we start directly with the inequality

$$|p\log x - q\log y| \ll \frac{1}{x^p}$$

whence arguing as before, shifting between y and $y+1$, we find:

$$|p\log x - q\log(q^{\delta'}Y^p)| \ll p^2/Y^p$$

and therefore

$$|p\log(x/Y^q) - q\delta'\log q| \ll p^2/Y^p = e^{-p\log Y + 2\log p}.$$

By the Baker–Tijdeman inequality, we get

$$C_0^{-(\log p)^{\theta+1}\log Y^q} \leqslant C_1\, e^{-p\log Y + 2\log p}.$$

By Step 1 we have already obtained a bound for q as a power of $\log p$. This time, we may cancel $\log Y$ in the inequality of the exponents, and thus end up with

$$p \ll q(\log p)^{\theta+1},$$

which proves what we wanted.

Observe that the Baker–Tijdeman inequality is applied here to linear combinations of rational numbers, with rational coefficients, and that $r = 2$ or 3.

The number theoretic trick at the beginning depended on having the constant 1 on the right-hand side of the Catalan equation. At the time this book is written, no proof has come to my attention which shows how to generalize the result to an equation

$$x^p - y^q = k$$

with arbitrary k.

It would be of tremendous interest to extend all of this to bound degrees of integral diophantine equations in contexts of algebraic geometry.

Chapter X. The Baker–Tijdeman Theorem

§ 1. Statement of the Theorem

Let $\alpha_1, \ldots, \alpha_r$ be algebraic numbers, in a number field K. We shall use the notation

$$u_j = \log \alpha_j ,$$

where the log is taken with principal value. We let

$$U_j = \log H_K(\alpha_j) \quad \text{and} \quad U = \max U_j ,$$

We do not assume u_j linearly independent over the integers. In the applications, we shall consider questions of uniformity only with respect to the degree of these algebraic numbers, and so we shall not pay special attention to distinguish the absolute from the relative height.

Lemma (Liouville inequality). *Let $b_1, \ldots, b_r$ be integers and let $B = \max b_j$. If $b_1 u_1 + \cdots + b_r u_r \neq 0$, then*

$$2^{-r-1} e^{-rBU} \leqslant |b_1 u_1 + \cdots + b_r u_r| .$$

Proof. We estimate the height of $\alpha_1^{b_1} \ldots \alpha_r^{b_r} - 1$. If v is non-archimedean then

$$|\alpha_1^{b_1} \ldots \alpha_r^{b_r} - 1|_v \leqslant \prod_{j=1}^{r} \max (1, |\alpha_j|_v)^B .$$

If v is archimedean, then we get the inequality

$$\leqslant \prod_{j=1}^{r} 2 \cdot \max (1, |\alpha_j|_v)^B .$$

Hence

$$H_K(\alpha_1^{b_1} \ldots \alpha_r^{b_r} - 1) \leqslant 2^r \prod_{j=1}^{r} H_K(\alpha_j)^B \leqslant 2^r e^{rBU} .$$

Hence we obtain

$$2^{-r} e^{-rBU} \leqslant |\alpha_1^{b_1} \dots \alpha_r^{b_r} - 1| .$$

Let $u = b_1 u_1 + \dots + b_r u_r$. If $|u| < \frac{1}{2}$, say, then u is the value of

$$\log (\alpha_1^{b_1} \dots \alpha_r^{b_r})$$

near 0, and we have with a generous estimate,

$$|e^u - 1| \leqslant 2|u| .$$

This proves the lemma.

Observe that instead of BU we could have put the better number

$$\sum_{j=1}^{r} B_j U_j ,$$

but this will not be required in the sequel.

For the rest of this chapter, we let $U_{n-1} = \max (U_1, \dots, U_{n-1})$,

$$\boxed{\tau(B, u) = (\log B) U_{n-1}^{\theta} U_n}$$

where θ is a number

$$\boxed{\theta > (2r + 3)(r + 3)}$$

and $r + 1$ is the maximum number of linearly independent elements among $u_1 \dots, u_n, 2\pi i$. Observe that $\theta > \kappa + 1$, where κ is the number of Chapter VIII. The precise form $(2r + 3)(r + 3)$ should be disregarded, treating all parameters as unknowns, and solving for necessary and sufficient conditions as we go along, terminating at the end of the last section with the above determination.

Theorem 1.1. *Let $\beta_1, \dots, \beta_n \in \mathbf{Q}$ and assume*

$$\beta_1 u_1 + \dots + \beta_n u_n \neq 0 .$$

Let $B = H_K(1, \beta_1, \dots, \beta_n)$. Then there is an inequality

$$|\beta_1 u_1 + \dots + \beta_n u_n| \geqslant C_0^{-\tau(B,u)}$$

where C_0 depends only on n and $[K : \mathbf{Q}]$.

As in the preceding chapter, we make a reduction to the case of linear independence. We again may assume $u_1 = 2\pi i$. Say $u_1, \ldots, u_r$ are maximal linearly independent among $u_1, \ldots, u_{n-1}$. If u_n depends on $u_1, \ldots, u_{n-1}$ then there is a rational relation

$$u_n = \sum_{\nu=1}^{r} c_\nu u_\nu \quad \text{with } H(c_\nu) \ll U^r \log\log U .$$

Substituting, we get

$$|\beta_1 u_1 + \cdots + \beta_n u_n| = |\beta_1' u_1 + \cdots + \beta_r' u_r| \leqslant C_0^{-\tau(B,u)}$$

and $B' = H_K(1, \beta_1', \ldots, \beta_r')$ satisfies

$$B' \ll BU^r \log\log U .$$

Therefore we have

$$(\log B') U_{n-1}^{\theta-1} \ll (\log B) U_{n-1}^{\theta} U_n = \tau(B, u) .$$

This contradicts the inequality proved in Chapter VIII.

Our function $\tau(B, u)$ will therefore be written in the form

$$\boxed{\tau(B, u) = (\log B) U_r^\theta U_{r+1} .}$$

We have assumed for convenience that $U_1 \leqslant U_2 \leqslant \cdots \leqslant U_r$.

If we divide the relation by β_n then we have the theorem in the most convenient form for proof:

Theorem 1.2. *Assume that $u_1, \ldots, u_{r+1}$ are linearly independent over the integers. Let $\beta_1, \ldots, \beta_r \in \mathbf{Q}$ and let $B = H_K(1, \beta_1, \ldots, \beta_r)$. Then there is an inequality*

$$|\beta_1 u_1 + \cdots + \beta_r u_r - u_{r+1}| \geqslant C_0^{-\tau(B,u)}$$

where C_0 is a constant depending only on r and $[K:\mathbf{Q}]$.

From now on, we assume the opposite inequality,

$$|\beta_1 u_1 + \cdots + \beta_r u_r - u_{r+1}| \leqslant C_0^{-\tau(B,u)}$$

and derive a contradiction.

Before proceeding further, we make two more reductions.

Reduction 1. We may assume that U_{r+1} is large compared to U_r, and specifically

$$U_{r+1} \geqslant 4rU_r .$$

For if $U_{r+1} \leqslant 4rU_r$ then $U \gg\ll U_r$ and the inequality contradicts the theorem of the last chapter since

$$\boxed{\theta \geqslant \kappa + 1 .}$$

Reduction 2. We may assume that for a sufficiently large constant C_1 we have

$$B \geqslant C_1 U_r^\theta ,$$

for if $B \leqslant C_1 U_r^\theta$ then the Liouville inequality yields

$$|\beta_1 u_1 + \cdots + \beta_r u_r - u_{r+1}| \geqslant C^{-U_r^\theta U_{r+1}}$$

which implies the theorem.

Remark. The use of the Liouville inequality is the first point where we make use of the fact that β_j are rational numbers rather than arbitrary algebraic numbers.

The second point will occur in connection with estimates of the Feldman polynomials and their derivatives, evaluated at rational numbers, rather than algebraic numbers.

§ 2. Main Lemma and its Application

We shall again work with parameters L, M, N. These will be of the following nature.

$$L_{r+1} = M_*^\sigma U_r^\rho$$

$$L = L_0 = \cdots = L_r = M_*^{\sigma'} U_r^{\rho'} U_{r+1}$$

$$M = M_* U_r^t U_{r+1}$$

$$N = [\log B] = L_{-1}$$

In these expressions, M_* is a sufficiently large constant, and $\sigma, \sigma', \rho, \rho', t$ are numbers > 0 for which explicit values will be given in the next section. What is essential here is that wherever it appears, U_{r+1} appears to the first power, and that $N = [\log B]$ also contains $\log B$ to the first power exactly. The other powers don't matter too much, and will be determined as functions of r. This indeterminacy reflects the fact that in the theorem we are trying to prove, any reasonable power of U_r is allowed in the approximating function $\tau(B, u)$.

We let

$$(\lambda) = (\lambda_{-1}, \lambda_0, \ldots, \lambda_{r+1}) \quad \text{and} \quad (m) = (m_0, \ldots, m_{r+1})$$

with

$$0 \leqslant \lambda_j \leqslant L_j, \quad j = -1, \ldots, r+1 .$$

We define

$$x_j(\lambda) = \lambda_j + \beta_j \lambda_{r+1}$$

and

$$\Psi_{\lambda,\, m}(x) = \Psi(x, \lambda, m, r) = \frac{1}{m_0!} D_0^{m_0} \Delta^{\lambda_0}(x + \lambda_{-1}, N) \prod_{j=1}^{r} \Delta(x_j(\lambda), m_j) .$$

We note that

$$\Delta^{\lambda_0}(x + \lambda_{-1}, N) \prod_{j=1}^{r} \Delta(x_j(\lambda), m_j)$$

is a polynomial in x of degree $\leqslant NL_0 = NL$. Its special form arises from the need to have such polynomials which take on as small values as possible when x is a rational number. In addition, the product on the right will allow us to perform an induction. We also observe that

$$x_j(\lambda)$$

arises from the exponents in

$$e^{z_j \lambda_j} e^{\beta_j z_j \lambda_j},$$

used in forming the approximating function.

Main Lemma. *For each M_* sufficiently large, there is a number C_0 having the following property. If there is an inequality*

$$|\beta_1 u_1 + \cdots + \beta_r u_r - u_{r+1}| \leqslant C_0^{-\tau(B,u)},$$

then there exist integer coefficients $a_{(\lambda)}$ not all 0, such that

$$(*) \qquad \sum_{(\lambda)} a_{(\lambda)} \Psi(n/q, m, \lambda, r) \alpha_1^{n\lambda_1/q} \ldots \alpha_{r+1}^{n\lambda_{r+1}/q} = 0$$

for some prime number q with $L_{r+1} < q < 2L_{r+1}$, and for all (m), satisfying

$$0 \leqslant m_j \leqslant L \quad and \quad n \leqslant M_* N .$$

The proof of the main lemma is postponed to the subsequent sections, and we proceed to show how it implies the theorem. As before, we wish to use the lemma in connection with Kummer theory. In the present case, an additional descent will be necessary because the original lemma is not strong enough to imply immediately that all the $a_{(\lambda)} = 0$. Thus we have to work with $\alpha_{r+1}^{n/q}$ separately. We shall also use the fact that instead of one polynomial equation, we have a large number of polynomial equations.

We view equations (*) as polynomial relations for $\alpha_{r+1}^{n/q}$ over the field

$$K_r = K(\alpha_1^{n/q}, \ldots, \alpha_r^{n/q}) .$$

We prove:

Step 1. *These polynomial relations are not all trivial, and hence $\alpha_{r+1}^{n/q}$ has degree $\leqslant L_{r+1} < q$ over K_r.*

Proof. We do this by induction on r. Suppose all relations are trivial. We rewrite (*) in the form

$$(**) \qquad \sum_{\lambda_r} \Delta_{\lambda_r, m_r} Y_{\lambda_r} = 0$$

where

$$\Delta_{\lambda_r, m_r} = \Delta(x_r(\lambda), m_r)$$

and

$$Y_{\lambda_r} = \sum_{\lambda_j, j \leqslant r-1} a_{(\lambda)} \Psi_{m, \lambda, r-1}\left(\frac{n}{q}\right) \alpha_1^{n\lambda_1/q} \ldots \alpha_r^{n\lambda_r/q} .$$

This is a system of linear equations, with determinant

$$\det_{\lambda_r, m_r} \Delta_{\lambda_r, m_r} \neq 0 .$$

Indeed, the polynomials

$$\Delta(x, 0), \Delta(x, 1), \ldots, \Delta(x, L)$$

are linearly independent by Lemma 2 of Chapter VII, § 4. If the determinant

$$\begin{vmatrix} \Delta(x_0, 0) & \Delta(x_0, 1) & \cdots & \Delta(x_0, L) \\ \vdots & \vdots & & \vdots \\ \Delta(x_L, 0) & \Delta(x_L, 1) & \cdots & \Delta(x_L, L) \end{vmatrix}$$

is zero for distinct numbers $x_0, x_1, \ldots, x_L$ there exists a non-trivial linear combination

$$w_0 \Delta(x, 0) + w_1 \Delta(x, 1) + \cdots + w_L \Delta(x, L)$$

which vanishes at $x_0, x_1, \ldots, x_L$, thus giving a contradiction. We conclude that for each λ_{r+1}, λ_r we have

$$\sum_{\lambda_j, j \leqslant r-1} a_{(\lambda)} \Psi_{m, \lambda, r-1} \left(\frac{n}{q}\right) \alpha_1^{n\lambda_1/q} \ldots \alpha_r^{n\lambda_r/q} = 0 .$$

Proceeding by induction we conclude that for each $m_0, n/q, \lambda_{r+1}, \ldots, \lambda_1$ there is a relation

$$\sum_{\lambda_{-1}, \lambda_0} a_{(\lambda)} \Delta\left(\frac{n}{q} + \lambda_{-1}, N, \lambda_0, m_0\right) = 0 .$$

Thus for each $\lambda_1, \ldots, \lambda_{r+1}$ the polynomial

$$\sum_{\lambda_{-1}, \lambda_0} a_{(\lambda)} \Delta(x + \lambda_{-1}, N)^{\lambda_0}$$

has zeros at all fractions n/q of multiplicity $\geqslant L$. The number of positive integers $n \leqslant M_* N$ relatively prime to q is at least

$$M_* N - M_* N/q \geqslant M_* N/2 .$$

Hence the above polynomial has at least

$$LM_* N/2$$

zeros, counting multiplicities, and this number is greater than its degree LN. This implies $a_{(\lambda)} = 0$ for all (λ), which is a contradiction, establishing the fact that $\alpha_{r+1}^{n/q}$ has degree $< q$ over K_r.

Step 2. By Kummer theory, there exists a multiplicative relation

$$\alpha_{r+1} \alpha_r^{\nu_1} \ldots \alpha_r^{\nu_r} = \alpha_{r+1}^{(1)q} , \quad 0 \leqslant \nu_j < q ,$$

for some non-zero element $\alpha_{r+1}^{(1)}$ in K. We can write this additively,

$$u_{r+1} + v_1 u_1 + \cdots + v_r u_r = q u_{r+1}^{(1)} + v \cdot 2\pi i\,,$$

with $|v| \ll q$. Also recall that $u_1 = 2\pi i$. From the definition of the height, we also have

$$H_K(\alpha_{r+1}^{(1)}) \leqslant H_K(\alpha_{r+1})^{1/q} H_K(\alpha_1) \ldots H_K(\alpha_r)\,,$$

whence putting $U_{r+1}^{(1)} = \log H_K(\alpha_{r+1}^{(1)})$ we get

$$U_{r+1}^{(1)} \leqslant \frac{1}{q} U_{r+1} + r U_r\,.$$

We substitute the above linear expression for u_{r+1} in the fundamental inequality of the main lemma. We then obtain

$$\begin{aligned} &|\beta_1 u_1 + \cdots + \beta_r u_r - u_{r+1}| \\ &= |(\beta_1 + v_1 - v) u_1 + (\beta_2 + v_2) u_2 + \cdots + (\beta_r + v_r) u_r - q u_{r+1}^{(1)}| \\ &= |\beta_1^{(1)} u_1 + \cdots + \beta_r^{(1)} u_r - q u_{r+1}^{(1)}| \\ &\leqslant C_0^{-\tau(B,u)}\,, \end{aligned}$$

where $\beta_j^{(1)}$ are new coefficients, satisfying

$$\text{Height of } \beta_j^{(1)} \leqslant B^{(1)} \quad \text{where} \quad B^{(1)} = B C_* U_r^{\rho}$$

and C_* depends only on M_* (and the usual r).

This new linear combination where u_{r+1} is replaced by $u_{r+1}^{(1)}$ has had the effect of increasing B to $B^{(1)}$ (the beginning of a geometric progression) but of decreasing U_{r+1} to $U_{r+1}^{(1)}$. To begin an induction, we show:

Under the reduction hypotheses

$$B \geqslant C_1 U_r^{\theta} \quad \textit{and} \quad U_{r+1} \geqslant 4 r U_r$$

it follows that

$$|\beta_1^{(1)} u_1 + \cdots + \beta_r^{(1)} u_r - q u_{r+1}^{(1)}| \;\leqslant C_0^{-\tau(B(1),\, u(1))}\,.$$

Proof. We need

$$(\log B^{(1)}) U_{r+1}^{(1)} \leqslant (\log B) U_{r+1}\,.$$

This is obvious because

$$(\log B + \log(C_* U_r^\rho))\left(\frac{1}{q} U_{r+1} + rU_r\right) \leqslant 2 \log B\left(\frac{1}{q} U_{r+1} + \frac{1}{4} U_{r+1}\right)$$
$$\leqslant (\log B) U_{r+1}\,.$$

This implies that we can perform this inductive step s times, obtaining coefficients $\beta_j^{(s)}$ satisfying an inequality

$$|\beta_1^{(s)} u_1 + \cdots + \beta_r^{(s)} u_r + \beta_{r+1}^{(s)} u_{r+1}^{(s)}| \leqslant C_0^{-\tau(B^{(s)}, u^{(s)})}$$

where

$$s \leqslant 2 \log U_{r+1}\,,$$

and Height $\beta_j^{(s)} \leqslant B^{(s)}$, as long as the reduction hypothesis is satisfied.

At the last step of this sequence, we have

$$U_{r+1}^{(s)} \leqslant 4rU_r\,.$$

and the final inequality contradicts the theorem of the preceding chapter provided $\theta \geqslant \kappa$. This concludes the proof.

§ 3. Construction of the System of Linear Equations

We begin by some remarks on the type of linear equations to be considered. Let P_{m_j} ($j = 1, \ldots, r$) be a polynomial of degree m_j, in one variable. Let

$$P_{(m)}(X_1, \ldots, X_r) = P_{m_1}(X_1) \cdots P_{m_r}(X_r)\,.$$

Let $P_{m_j}^*$ be another polynomial of degree m_j. The family

$$\{P_{(m)}(X_1, \ldots, X_r)\} \quad \text{for} \quad |m| \leqslant M$$

forms a basis for the space of polynomials of degree $\leqslant M$, and similarly for $\{P_{(m)}^*\}$. Consequently, keeping the notation of the last section,

$$x_j(\lambda) = \lambda_j + \beta_j \lambda_{r+1}$$

and letting

$$\psi_{m_0, \lambda_0, \lambda_{-1}}(z_0), f_1(z_1), \ldots, f_r(z_r)$$

be functions of one variable, we see that solving the system of linear equations for certain values of z:

$$\begin{aligned}\mathrm{Equ}\,(m, z) &= \sum a_{(\lambda)}\psi_{m_0,\lambda_0,\lambda_{-1}}(z)\,P_{(m)}(x_1(\lambda), \ldots, x_r(\lambda))f_1(z)^{\lambda_1}\cdots f_r(z)^{\lambda_r}\\ &= 0 \quad \text{for all } |m| \leqslant M\end{aligned}$$

is equivalent to solving the system

$$\begin{aligned}\mathrm{Equ}^*\,(m, z) &= \sum a_{(\lambda)}\psi_{m_0,\lambda_0,\lambda_{-1}}(z)\,P^*_{(m)}(x_1(\lambda), \ldots, x_r(\lambda))f_1(z)^{\lambda_1}\cdots f_r(z)^{\lambda_r}\\ &= 0 \quad \text{for all } |m| \leqslant M\,.\end{aligned}$$

The ordinary system of monomials $\{X_1^{m_1}\ldots X_r^{m_r}\}$ is of course the most common, but in order to get a better estimate for polynomials evaluated at rational numbers, we shall choose instead the system

$$\{\Delta(X_j, m_j\} \quad \text{with } |m| \leqslant M\,.$$

As in the preceding section, we let

$$\Psi_{m,\lambda}(x) = \frac{1}{m_0!}\,D_0^{m_0}\Delta^{\lambda_0}(x + \lambda_{-1}, N)\prod_{j=1}^{r}\Delta(\lambda_j + \beta_j\lambda_{r+1}, m_j)\,.$$

We also repeat the determination of parameters:

$$\begin{aligned} N &= [\log B] = L_{-1}\\ L_{r+1} &= M_*^{\sigma}U_r^{\varrho}\\ L = L_0 = \cdots = L_r &= M_*^{\sigma'}U_r^{\varrho'}U_{r+1}\\ M &= M_*U_r^tU_{r+1}\end{aligned}$$

We wish to solve the system of linear equations for $a_{(\lambda)}$:

$$\mathrm{Equ}\,(m, n) = \sum a_{(\lambda)}\,\Psi_{m,\lambda}(n)\,e^{nu_1\lambda_1}\ldots e^{nu_r\lambda_r} = 0$$

for

$$|m| \leqslant M \quad \text{and} \quad 1 \leqslant n \leqslant N\,.$$

We have:

$$\text{number of variables} \geqslant L_{-1}L^{r+1}L_{r+1} = NL^{r+1}L_{r+1}$$

$$\text{number of equations} \leqslant M^{r+1}N\,.$$

We want

$$\text{number of variables} \geqslant 2[K:\mathbf{Q}]\ \text{number of equations},$$

so we pick the cheapest way of achieving

$$L^{r+1}L_{r+1} \geqslant 2[K:\mathbf{Q}]M^{r+1} = 2[K:\mathbf{Q}]M_*^{r+1}U_r^{t(r+1)}U_{r+1}^{r+1}\,. \tag{1}$$

We also have

$$\text{height of coefficients} \leqslant C^{LN}C^{MN}C^{(L+L_{r+1}+M)}\,e^{\sum NU_jL_j}B^M\,.$$

because of the estimates for the Feldman polynomials given in Chapter VII, § 4. Observe that we use here the fact that β^j is rational and not algebraic.

As before, if $X = M_*^a U_r^b X'$ and $Y = M_*^{a'} U_r^{b'} Y'$ where X', Y' do not contain M_* or U_r, we use the notation

$$X \prec Y \quad \text{to mean } a < a' \text{ and } b < b'\,.$$

We shall want

$$\boxed{U_jL_j \prec M \quad \text{for } j = 1, \ldots, r+1\,.} \tag{2}$$

If this condition is satisfied, then we see that

$$\text{height of coefficients} \leqslant C^{MN}.$$

All of this is with an appropriate constant C. It will turn out that all further conditions we need will be subsumed by (1) and (2). We must now see that we can solve for $\sigma, \sigma', \rho, \rho'$ to satisfy these conditions. We deal with the M_* factor and the U_r factor separately.

M_* factor. For (1), we need $(r+1)\sigma + \sigma' \geqslant r+1$.
For (2), we need $\sigma, \sigma' < 1$.

One can solve these for instance with

$$\boxed{\sigma = 1 - \frac{1}{4(r+1)} \quad \text{and} \quad \sigma' = \frac{1}{2}}$$

but anything else would do also.

U_r **factor.** For (1) we need $(r+1)\rho + \rho' \geqslant t(r+1)$
For (2) we need $\rho + 1 < t$ and $\rho' < t$.

We can take as solution

$$\boxed{\rho = t - 2 \quad \text{and} \quad \rho' = t - 1}$$

in which case this works if $t \geqslant 2r + 3$, so we pick

$$\boxed{t = 2r + 3\,.}$$

This settles our choice of parameters since we can then solve for L, L_{r+1}.

In the light of the remarks made at the beginning of the section, we see that:

the system of equations

$$\text{Equ}\,(m, n) = 0, \quad |m| \leqslant M \text{ and } 1 \leqslant n \leqslant N$$

implies

$$\sum a_{(\lambda)} D_0^{m_0} \Delta^{\lambda_0}(n + \lambda_{-1}, N) P^*(x_1(\lambda), \ldots, x_r(\lambda)) \alpha_1^{n\lambda_1} \ldots \alpha_{r+1}^{n\lambda_{r+1}} = 0$$

where P^* is any polynomial of total degree $\leqslant M$ and again $1 \leqslant n \leqslant N$.

§ 4. Extrapolation on Integral Multiples

Lemma 4.1. *There exists δ with $0 < \delta < 1$ having the following property. Let*

$$S_\nu = N(M_* U_r^t)^{\nu\delta}.$$

Then the system of linear equations

$$\text{Equ}\,(m, n) = 0$$

is satisfied for

$$n \leqslant S_\nu \quad \textit{and} \quad |m| \leqslant M/2^\nu,$$

for all integers ν such that $\nu\delta \leqslant r + 2$.

Proof. By induction. We assume the lemma for ν and prove it for $\nu + 1$. Let

$|m'| \leqslant M/2^{\nu+1}$. Let

$$G_{m'}(z_0, \ldots, z_r) = \sum a_{(\lambda)} \Psi_{m',\lambda}(z_0)\, e^{z_1\lambda_1} \ldots e^{z_r\lambda_r}\, e^{(\beta_1 z_1 + \cdots + \beta_r z_r)\lambda_{r+1}} .$$

Then for $(k) = (k_0, \ldots, k_r)$ *such that* $|k| \leqslant M/2^{\nu+1}$ *and* $n \leqslant S_\nu$ *we have*

$$D^{(k)} G_{m'}(n, nu_1, \ldots, nu_r; nu_{r+1}) = 0 .$$

Indeed, this expression is of the form

$$\sum a_{(\lambda)} D_0^{m'_0+k_0} \Delta^{\lambda_0}(n + \lambda_{-1}, N) \prod_{j=1}^{r} \Delta(x_j(\lambda), m'_j) \prod_{j=1}^{r} x_j(\lambda)^{k_j}\, \alpha^{(n\lambda)}$$

where $\alpha^{(n\lambda)}$ is the usual vector abbreviation,

$$\alpha^{(n\lambda)} = \alpha_1^{n\lambda_1} \ldots \alpha_{r+1}^{n\lambda_{r+1}} .$$

The two products over $j = 1, \ldots, r$ constitute a polynomial in $x_1(\lambda), \ldots, x_r(\lambda)$ of total degree $\leqslant M/2$, and our assertion follows from the remark made at the end of the last section and the induction hypothesis.

We now need an estimate for the difference of derivatives and substituted derivatives as in the preceding chapter.

Lemma 4.2. *Let* $N \leqslant S$ *and suppose* $SM \leqslant \tau(B, u)$. *Then for* C_0 *sufficiently large and* $n \leqslant S$ *we have*

$$\left| \frac{1}{k_0!} D^{(k)} G_{m'}(n, nu_1, \ldots, nu_r) - \frac{1}{k_0!} D^{(k)} G_{m'}(n, nu_1, \ldots, nu_r; nu_{r+1}) \right| \leqslant C_0^{-\tau(B,u)/2} .$$

Proof. It is easy to estimate a derivative of $G_{m'}$. We note that

$$\frac{1}{k_0!} D_0^{(k)} G_{m'}(n, nu_1, \ldots, nu_r)$$

as a polynomial consists of three types of terms.

A term with

$$\frac{1}{k_0! m'_0!} D_0^{m'_0+k_0} \Delta^{\lambda_0}(n + \lambda_{-1}, N)$$

estimated by C^{SL}, with a denominator $\leqslant C^{MN}$ by Chapter VII, § 4.

Terms similar to those of § 3, and already estimated by C^{MN}.

Monomials involving the expanded range for n,

$$e^{nu_1\lambda_1} \ldots e^{nu_r\lambda_r}\, e^{n(\beta_1 u_1 + \cdots + \beta_r u_r)\lambda_{r+1}} \ll C^{S\sum U_j L_j} .$$

Since substituting u_{r+1} for $\beta_1 u_1 + \cdots + \beta_r u_r$ creates a change estimated by

$$C_0^{-\tau(B,u)},$$

we see that the lemma is proved.

We apply Lemma 2 to the case $S = S_\nu$. Since we have seen that the substituted derivative is equal to 0, that is

$$\frac{1}{k_0!} D^{(k)} G_{m'}(n, nu_1, \ldots, nu_r; nu_{r+1}) = 0 ,$$

it follows that the actual derivative is small, that is

$$\left| \frac{1}{k_0!} D^{(k)} G_{m'}(n, nu_1, \ldots, nu_r) \right| \leqslant C_0^{-\tau(B,u)/2},$$

for $n \leqslant S_\nu$.

We are now in a position to apply the Hermite interpolation formula, to the function of one variable

$$\begin{aligned} g(z) &= \sum a_{(\lambda)} \Psi_{m',\lambda}(z)\, e^{u_1 z \lambda_1} \cdots e^{u_r z \lambda_r}\, e^{(\beta_1 u_1 + \cdots + \beta_r u_r) z \lambda_{r+1}} \\ &= G_{m'}(z, u_1 z, \ldots, u_r z) . \end{aligned}$$

We pick ε small (to satisfy (4) below) and let the radius R be

$$R = S_{\nu+1} M_*^{\varepsilon} .$$

We are trying to show that the equations $\mathrm{Equ}(m', n') = 0$ are satisfied for $n' \leqslant S_{\nu+1}$. We therefore estimate $g(n')$. We find:

$$|g(n')| \leqslant \frac{\|g\|_R C^{MS_\nu}}{C^{MS_\nu \log M_*}} + (CS_{\nu+1}/S_\nu)^{MS_\nu} \max \left| \frac{1}{k_0!} D^{(k)} g(n) \right|$$

where the maximum of the derivatives is taken for $n \leqslant S_\nu$. We have written a constant C three times, for simplicity. These constants should be indexed consecutively. We have

$$S_{\nu+1}/S_\nu = (M_* U_r^t)^\delta .$$

Therefore the second term on the right will be small provided that

$$(3) \qquad \boxed{MS_\nu < \tau(B,u) \quad \text{or equivalently} \quad (\nu\delta + 1)t < \theta .}$$

As to the first term, we have the easy estimate using (2):

$$\|g\|_R \leqslant C^{R\sum U_j L_j}.$$

Therefore the denominator wins over the numerator provided that

$$\text{(4)} \qquad \boxed{S_{\nu+1} M_*^{\varepsilon}\, U_j L_j < M S_\nu \quad \text{or equivalently} \quad M_*^{\varepsilon} (M_* U_r^!)^{\delta}\, U_j L_j < M\,.}$$

Since we had $U_j L_j < M$ by (2), we can find ε, δ small enough to satisfy this condition (4) also.

We may therefore conclude that there is a constant C_1 depending on the usual r, $[K:\mathbf{Q}]$ such that for M_* large,

$$|g(n')| \leqslant \frac{1}{C_1^{MS_\nu \log M_*}}.$$

By Lemma 2, the algebraic number $\xi = G_{m'}(n', n'u_1, \ldots, n'u_r; n'u_{r+1})$ is within $C_0^{-\tau(B,u)/2}$ of $g(n')$, and consequently by (3) this algebraic number in K satisfies the same sort of upper estimate. Since n' ranges to $S_{\nu+1}$ we also get a lower bound for its absolute value in terms of its size, if it is not zero, namely

$$\frac{1}{C_2^{\sum U_j L_j S_{\nu+1}}} \leqslant |\xi| \leqslant \frac{1}{C_3^{S_\nu M \log M_*}}.$$

This is a contradiction by (4), which shows that

$$0 = \xi = G_{m'}(n', n'u_1, \ldots, n'u_r; n'u_{r+1}) = \operatorname{Equ}(m', n'),$$

and proves Lemma 1.

§ 5. Extrapolation on Fractional Multiples

Main Lemma. *The system of equations*

$$\operatorname{Equ}\left(m, \frac{n}{q}\right) = 0$$

is satisfied for

$$n \leqslant NM_*, \qquad 0 \leqslant m_j \leqslant L$$

and all integers q *with* $2 < q \leqslant 2L_{r+1}$.

Proof. The steps are entirely similar to those of the last section, except at the very last estimate. Let us abbreviate

$$Y = NM_*,$$

which gives the range for n in the lemma. We let $m_j' \leqslant L$ and let $n' \leqslant T$. We end up with the algebraic number

$$\xi = G_{m'}\left(\frac{n'}{q}, \frac{n'}{q}u_1, \ldots, \frac{n'}{q}u_r; \frac{n'}{q}u_{r+1}\right)$$

by the same construction as before, satisfying the estimates

$$\frac{1}{C_2^{\sum U_j L_j T q^{r+1}}} \leqslant |\xi| \leqslant \frac{1}{C_3^{SM}}$$

where S is the maximum value S_v allowed in Lemma 1 of § 4. In the present case, the lower estimate involving the height has the additional q^{r+1} because the height has to be taken with respect to a field of degree $\leqslant q^{r+1}$ over K, obtained by extracting q-th roots of $\alpha_1, \ldots, \alpha_{r+1}$.

We then get a contradiction to $\xi \neq 0$ if

(5) $$\boxed{U_j L_j T q^{r+1} \prec MS},$$

for which it suffices that

(6) $$NM_*(M_*^\sigma U_r^\rho)^{r+1} \prec S.$$

This step shows us how far to push the induction on v in § 4, namely so far that

$$M_*(M_*^\sigma U_r^\rho)^{r+1} \prec (M_* U_r^t)^{v\delta}.$$

Using the values for σ, ρ, t already determined, we see that

$$v\delta = r + 2$$

will do.

The application of the Hermite formula had also led us to require (3), namely

$$MS \prec \tau(B, u) \quad \text{or equivalently} \quad (v\delta + 1)t < \theta.$$

We now know the maximal value of $v\delta$, so that this condition reads

$$\theta > (r + 3)(2r + 3),$$

which concludes the proof.

Chapter XI. Refined Inequalities

§ 1. Statement of the Theorem

Again we let $\alpha_1, \ldots, \alpha_r$ be algebraic numbers in a number field K. We let

$$u_j = \log \alpha_j$$

and

$$U_j = \log H_K(\alpha_j)\,.$$

We do not assume u_j linearly independent over the integers. We also assume

$$U_{r-1} = \max(U_1, \ldots, U_{r-1})\,.$$

For the rest of this chapter, we let

$$\boxed{\tau(B, u) = (\log B) U_1 \cdots U_r \log U_{r-1}\,.}$$

Here we give the proof of the following result of Van der Poorten [VdP 1] and Baker [Ba 7] (see also [VdP–L 1], [VdP–L 2]). I am much indebted to Waldschmidt for co-authoring the rest of this chapter.

The proof is self contained except for the simple Liouville inequality in § 1 of Chapter X.

Theorem 1.1. *Let $\beta_1, \ldots, \beta_r$ be rational, and assume*

$$\beta_1 u_1 + \cdots + \beta_r u_r \neq 0\,.$$

Let $B = H_{\mathbf{Q}}(1, \beta_1, \ldots, \beta_r)$. Then there is an inequality

$$|\beta_1 u_1 + \cdots + \beta_r u_r| \geqq C_0^{-\tau(B, u)}\,,$$

where C_0 depends only on r and $[K : \mathbf{Q}]$.

As in the preceding chapters, we shall make a reduction to a case of independent $\alpha_1, \ldots, \alpha_r$. But here this reduction will be done in connection with Kummer's theory, and we postpone it to § 5. The main part of this chapter will be devoted to the proof of the following special case of Theorem 1.1.

Theorem 1.2. *Assume* $[K(\alpha_1^{\frac{1}{2}}, \ldots, \alpha_r^{\frac{1}{2}}) : K] = 2^r$. *Let* $\beta_1, \ldots, \beta_{r-1} \in \mathbf{Q}$, *and let* $B = H_{\mathbf{Q}}(1, \beta_1, \ldots, \beta_{r-1})$. *Then there is an inequality*

$$|\beta_1 u_1 + \cdots + \beta_{r-1} u_{r-1} - u_r| \geqslant C_0^{-\tau(B,u)}.$$

From now on we assume the opposite inequality

$$|\beta_1 u_1 + \cdots + \beta_{r-1} u_{r-1} - u_r| \leqslant C_0^{-\tau(B,u)},$$

and derive a contradiction at the end of § 4.

Our theorems hold for $B \geqslant 2$, but it is sufficient to prove them for large B, for otherwise we replace B by B^C, with large C, and we get the result with C_0 replaced by C_0^C:

$$C_0^{-\tau(B^C,u)} = (C_0^C)^{-\tau(B,u)}.$$

For the same reason we assume that $U_1, \ldots, U_r$ are large in comparison to $[K : \mathbf{Q}]$ and r.

§ 2. Main Lemma and its Application

We shall again work with parameters N, L, M, namely

$$\begin{aligned}
N &= L_{-1} = [\log B] \\
L_0 &= M_*^{r+1} U_1 \ldots U_r \\
L_j &= M_*^r U_1 \ldots U_{j-1} U_{j+1} \ldots U_r \log U_{r-1}, \quad (1 < j \leqslant r) \\
M &= M_*^{r+1} U_1 \ldots U_r \log U_{r-1}.
\end{aligned}$$

In these expressions, M_* is a sufficiently large constant that can be specified in terms of r and $[K : \mathbf{Q}]$ only. Our choice will be explained later on (end of § 4); what is essential here is that whenever it appears, U_r appears to the first power, and that N also contains $\log B$ to the first power exactly.

Without loss of generality, we may assume that U_j and B are sufficiently large in comparison to M_*.

We let $(m) = (m_0, \ldots, m_{r-1})$, and, for any non negative integer J,

$$(\lambda) = (\lambda^{(J)}) = (\lambda_1, \ldots, \lambda_r),$$

with

$$0 \leqslant \lambda_{-1} \leqslant L_{-1}, \qquad 0 \leqslant \lambda_0 \leqslant L_0 \, ,$$

and

$$0 \leqslant \lambda_j \leqslant L_j/2^J, \qquad (j = 1, \dots, r).$$

We define

$$x_j(\lambda) = \lambda_j + \beta_j \lambda_r$$

and

$$\Psi_{\lambda, m}(x) = \Psi(x, \lambda, m, r-1) = \frac{1}{m_0!} D_0^{m_0} \Delta^{\lambda_0}(x + \lambda_{-1}, L_{-1}) \prod_{j=1}^{r-1} \Delta(x_j(\lambda), m_j) \, .$$

We note that

$$\Delta^{\lambda_0}(x + \lambda_{-1}, L_{-1}) \prod_{j=1}^{r-1} \Delta(x_j(\lambda), m_j)$$

is a polynomial in x of degree $\leqslant L_{-1} L_0$. Its special form arises from the need to have such polynomials which take on as small values as possible when x is a rational number. In addition, the product on the right will allow us to perform an induction. We also observe that

$$x_j(\lambda)$$

arises from the exponent in

$$e^{\lambda_j z_j} e^{\lambda_r \beta_j z_j},$$

used in forming the approximating function.

Main lemma. *For each M_* sufficiently large, there is a number C_0 having the following property. If there is an inequality*

$$|\beta_1 u_1 + \cdots + \beta_{r-1} u_{r-1} - u_r| \leqslant C_0^{-\tau(B,u)}$$

with the hypotheses of Theorem 1.2, then for each integer $J \geqslant 0$ satisfying

$$2^{J-1} \leqslant L_r \, ,$$

there exist integers $a^{(J)}(\lambda^{(J)})$ not all zero, such that

$$\sum_{(\lambda^{(J)})} a^{(J)}(\lambda^{(J)})\Psi\left(\frac{n}{2^J}, m, \lambda^{(J)}, r-1\right)\alpha_1^{n\lambda_1}\ldots\alpha_r^{n\lambda_r} = 0$$

for all (m) satisfying $|m| \leqslant M/2^J$ and all odd n satisfying $1 \leqslant n \leqslant 2^J N$.

The proof of the main lemma is postponed to the subsequent sections, and we proceed to show how it implies the theorem. We write the conclusion of the main lemma for

$$J = J_0 = \left[\frac{\log L_r}{\log 2}\right] + 1\,.$$

Thus we have

$$2^{J_0} > L_r\,,$$

hence $[L_r/2^{J_0}] = 0$. We conclude that for all (m) satisfying

$$0 \leqslant m_0 \leqslant M/2^{J_0+1}\,, \qquad 0 \leqslant m_j \leqslant L_j/2^{J_0} \qquad (1 \leqslant j \leqslant r-1)$$

and all odd n satisfying $1 \leqslant n \leqslant 2^{J_0}N$, we have

$$\sum_{\lambda_{r-1}} \Delta(\lambda_{r-1}, m_{r-1})Y_{\lambda_{r-1}} = 0\,,$$

where λ_{r-1} runs over $0 \leqslant \lambda_{r-1} \leqslant L_{r-1}/2^{J_0}$, and

$$Y_{\lambda_{r-1}} = \sum_{\lambda_j, j \leqslant r-2} a^{(J_0)}(\lambda^{(J_0)})\Psi\left(\frac{n}{2^{J_0}}, m, \lambda^{(J_0)}, r-2\right)\alpha_1^{n\lambda_1}\ldots\alpha_{r-2}^{n\lambda_{r-2}}\,.$$

This is a system of linear equations in $Y_1, \ldots, Y_L$ (with $L = [L_{r-1}/2^{J_0}]$), with determinant

$$\det_{\lambda_{r-1}, m_{r-1}} \Delta(\lambda_{r-1}, m_{r-1}) \neq 0\,.$$

Indeed the polynomials

$$\Delta(x, 0), \Delta(x, 1), \ldots, \Delta(x, L)$$

are linearly independent by Lemma 2 of Chapter VII, § 4; if the determinant

$$\begin{vmatrix} \Delta(x_0, 0) & \Delta(x_0, 1) & \ldots & \Delta(x_0, L) \\ \vdots & \vdots & & \vdots \\ \Delta(x_L, 0) & \Delta(x_L, 1) & \ldots & \Delta(x_L, L) \end{vmatrix}$$

is zero for distinct numbers $x_0, \ldots, x_L$ there exists a non-trivial linear combination

$$w_0 \Delta(x, 0) + w_1 \Delta(x, 1) + \cdots + w_L \Delta(x, L)$$

which vanishes at $x_0, x_1, \ldots, x_L$, thus giving a contradiction. We conclude that

$$Y_1 = Y_2 = \cdots = Y_L = 0\,,$$

which means that for each λ_{r-1}, we have

$$\sum_{\lambda_j,\, j \leqslant r-2} a^{(J_0)}(\lambda^{(J_0)}) \Psi\left(\frac{n}{2^{J_0}}, m, \lambda^{(J_0)}, r-2\right) \alpha_1^{n\lambda_1} \ldots \alpha_{r-2}^{n\lambda_{r-2}} = 0\,.$$

Proceeding by induction we conclude that for each $m_0, n, \lambda_{r-1}, \ldots, \lambda_1$ there is a relation

$$\sum_{\lambda_{-1}, \lambda_0} a^{(J_0)}(\lambda^{(J_0)}) D_0^{m_0} \Delta^{\lambda_0}\left(\frac{n}{2^{J_0}} + \lambda_{-1}, L_{-1}\right) = 0\,.$$

Thus for each $\lambda_1, \ldots, \lambda_{r-1}$, the polynomial

$$\sum_{\lambda_{-1}, \lambda_0} a^{(J_0)}(\lambda^{(J_0)}) \Delta^{\lambda_0}(x + \lambda_{-1}, L_{-1})$$

has at least

$$2^{J_0-1} N \cdot [M/2^{J_0+1}]$$

zeros, counting multiplicities, and this number is greater than its degree $L_{-1}L_0$. This implies $a^{(J_0)}(\lambda^{(J_0)}) = 0$ for all $(\lambda^{(J_0)})$, which is a contradiction, establishing Theorem 1.2.

Observe that in the main lemma, the number of equations, i.e. the number of $((m), r)$, is independent of J, roughly $M^r N$, while the number of coefficients, i.e. the number of $(\lambda^{(J)})$, is a decreasing function of J. The comparison between these two numbers allows us to use Siegel's lemma for $J = 0$, and gives a contradiction for J large. Usually (for instance in the preceding chapters), the number of coefficients is fixed, and the number of equations increases along the induction.

§ 3. Construction of the System of Linear Equations

We begin by some remarks on the type of linear equations to be considered. For $1 \leqslant j \leqslant r-1$ and $m_j \geqslant 0$, let P_{m_j} be a polynomial of exact degree m_j, in one variable.

Let

$$P_{(m)}(X_1, \ldots, X_{r-1}) = P_{m_1}(X_1) \ldots P_{m_{r-1}}(X_{r-1}).$$

Let P^*_{mj} be another polynomial of degree m_j, in one variable. The family

$$\{P_{(m)}(X_1, \ldots, X_{n-1})\} \quad \text{for } |m| \leqslant M$$

forms a basis for the space of polynomials of degree $\leqslant M$, and similarly for $P^*_{(m)}$. Consequently, keeping the notation of the 2nd section,

$$x_j(\lambda) = \lambda_j + \beta_j \lambda_n,$$

and letting

$$\theta_{m_0, \lambda_0, \lambda-1}(z_0), f_1(z_1), \ldots, f_r(z_r)$$

be functions of one variable, we see that solving the system of linear equations for certain value of z:

Equ(m, z) *for all* $|m| \leqslant M$:

$$\sum_{(\lambda)} a(\lambda) \theta_{m_0, \lambda_0, \lambda-1}(z) P_{(m)}(x_1(\lambda), \ldots, x_{r-1}(\lambda)) f_1(z)^{\lambda_1} \ldots f_r(z)^{\lambda_r} = 0$$

is equivalent to solving the system Equ*$(m, z) = 0$ obtained by replacing $P_{(m)}$ by $P^*_{(m)}$.

The ordinary system of monomials $x_1^{m_1} \ldots x_{r-1}^{m_{r-1}}$ is of course the most common, but in order to get a better estimate for polynomials evaluated at rational numbers, we shall choose instead the system

$$\{\Delta(X_j, m_j)\} \quad \text{with} \quad |m| \leqslant M.$$

As in the preceding section, we let

$$\Psi_{\lambda, m}(x) = \frac{1}{m_0!} D_0^{m_0} \Delta^{\lambda_0}(x + \lambda_{-1}, L_{-1}) \prod_{j=1}^{r-1} \Delta(x_j(\lambda), m_j).$$

In this section we write (λ) for $(\lambda^{(0)})$. We wish to solve the system of linear equations for $a(\lambda)$:

$$\textbf{Equ}(\boldsymbol{m}, \boldsymbol{n}) = \sum_{(\lambda)} a(\lambda) \Psi_{\lambda, m}(n)\, e^{nu_1 \lambda_1} \cdots e^{nu_r \lambda_r} = 0$$

for

$$|m| \leqslant M \quad \text{and} \quad 1 \leqslant n \leqslant N .$$

We have:

$$\text{number of variables } \geqslant L_{-1} L_0 \ldots L_r$$

$$\text{number of equations } \leqslant (M+1)^r N .$$

Our choice of parameters leads to

$$\text{number of variables} \geqslant 2[K:\mathbf{Q}] \cdot \text{number of equations}.$$

We also have an upper bound for the size of the matrix of coefficients:

$$\text{size (coefficients)} \leqslant C^{L_{-1} M} B^{CM} \prod_{j=1}^{r} C^{NL_j U_j} C^{L_0 (H+L_{-1})} B^{CL_r},$$

because of the estimates for the Feldman polynomials given in Chapter VII, § 4.

Observe that we use here the fact that β_j is rational: when β is, say, an algebraic integer, the number

$$\Delta(\beta, k),$$

(for $k \geqslant 1$, k integer) may have a large denominator.

Using Lemma 2 of Chapter VII § 2 (Siegel's lemma), we obtain a non trivial solution $a(\lambda)$ in $\mathbf{Z}$, with

$$\log \max_{(\lambda)} |a(\lambda)| \leqslant CMN .$$

In the light of the remarks made at the beginning of the section, we see that *the system of equations*

$$\text{Equ}\,(m, n) = 0, \quad |m| \leqslant M \text{ and } 1 \leqslant n \leqslant N$$

implies

$$\sum a(\lambda) D_0^{m_0} \Delta^{\lambda_0}(h + \lambda_{-1}, L_{-1}) P^*(x_1(\lambda), \ldots, x_{n-1}(\lambda)) \alpha_1^{n\lambda_1} \ldots \alpha_r^{n\lambda_r} = 0$$

when P^* is any polynomial of total degree $\leqslant M$ and again $1 \leqslant n \leqslant N$.

We have written a constant C several times, for simplicity, and we shall do the same in the next section. These constants should be indexed consecutively, and we leave it to the reader.

§ 4. Proof of the Main Lemma

We prove the main lemma of § 2 by induction on J. For $J = 0$ we choose

$$a^{(0)}(\lambda^{(0)}) = a(\lambda)\,,$$

thanks to the preceding section; we recall that

$$\log \max_{(\lambda)} |a(\lambda)| \leqslant CMN\,.$$

We assume that the assertion of the main lemma is correct for some integer J with $1 \leqslant 2^J \leqslant L_r$, and with

$$\log \max |a^{(J)}(\lambda^{(J)})| \leqslant CMN\,,$$

and we prove it for $J+1$. From now on (λ) stands for $(\lambda^{(J)})$.

Let $|m'| \leqslant M/2^{J+1}$; let

$$G_{m'}(z_0, \ldots, z_{r-1}) = \sum a^{(J)}(\lambda^{(J)})\Psi_{\lambda, m'}\left(\frac{z_0}{2^J}\right) e^{z_1\lambda_1} \ldots e^{z_{r-1}\lambda_{r-1}}\, e^{(\beta_1 z_1 + \cdots + \beta_{r-1} z_{r-1})\lambda_r}\,.$$

Then for $(k) = (k_0, k_1, \ldots, k_{n-1})$ *such that* $|k| \leqslant M/2^{J+1}$ *and* n *odd*, $1 \leqslant n \leqslant 2^J N$, *we have*

$$D^{(k)} G_{m'}(n; nu_1, \ldots, nu_{r-1}; nu_r) = 0\,.$$

Indeed this expression is of the form

$$\sum a^{(J)}(\lambda^{(J)}) D_0^{m'_0 + k} \Delta^{\lambda_0}\left(\frac{n}{2^J} + \lambda_{-1}, L_{-1}\right) \prod_{j=1}^{r-1} \Delta(x_j(\lambda), m'_j) \prod_{j=1}^{r-1} x_j(\lambda)^{k_j} \cdot \alpha^{(n,\,\lambda)}$$

where $\alpha^{(n\cdot\lambda)}$ is the usual vector abbreviation,

$$\alpha^{(n\cdot\lambda)} = \alpha_1^{n\lambda_1} \ldots \alpha_r^{n\lambda_r}\,.$$

The two products over $j = 1, \ldots, r-1$ constitute a polynomial in $x_1(\lambda), \ldots, x_{r-1}(\lambda)$ of total degree $\leqslant M$, and our assertion follows from the remark made at the end of the last section and the induction hypothesis.

We now need an estimate for the difference of derivatives and substituted derivatives as in the preceding chapter.

Lemma 4.1. *For* $|k| \leqslant M/2^{J+1}$, *and for complex* x *with* $|x| \leqslant 2^{J+1}N$, *we have*

$$|D^{(k)}G_{m'}(x, xu_1, \ldots, xu_{r-1}) - D^{(k)}\, G_{m'}(x, xu_1, \ldots, xu_{r-1}\,; xu_r)| \leqslant C_0^{-\tau(B,u)}$$

provided that C_0 *is sufficiently large (say* $C_0 = \exp\,(M_*^{r+2})$).

Proof. It is easy to estimate a derivative of $G_{m'}$. We note that

$$D_0^{(k)}G_{m'}(x, xu_1, \ldots, xu_{r-1})$$

consists of three types of terms.

A term with

$$\frac{1}{m_0!}\, D_0^{m_0'+k_0}\Delta^{\lambda_0}\left(\frac{x}{2^J} + \lambda_{-1}, L_{-1}\right)$$

estimated by C^{MN} by Chapter VII, § 4.

Terms similar to those of § 3, and already estimated by C^{MN}; we replace M by $M/2^J$, N by $2^{J+1}N$, and L_j by $L_j/2^J$ for $1 \leqslant j \leqslant r$; up to a factor 2, the terms NL_jU_j and NM are invariant.

Terms involving the expanded range for x,

$$e^{xu_1\lambda_1}\ldots e^{xu_{r-1}\lambda_{r-1}}\, e^{x(\beta_1 u_1+\cdots+\beta_{r-1}u_{r-1})\lambda_r}$$

whose absolute value is at most

$$C^{|x|\sum U_jL_j/2^J}.$$

Since substituting u_r for $\beta_1 u_1 + \cdots + \beta_{r-1}u_{r-1}$ creates a change estimated by

$$C_0^{-\tau(B,u)},$$

we see that the lemma is proved.

We apply Lemma 4.1 to the case $x = 2n$, with n odd, $1 \leqslant n \leqslant 2^JN$. Since we have seen that the substituted derivative is equal to zero, that is

$$D^{(k)}G_{m'}(n, nu_1, \ldots, nu_{r-1}\,; nu_r) = 0\,,$$

it follows that the actual derivative is small, that is

$$|D^{(k)}G_{m'}(n, nu_1, \ldots, nu_{r-1})| \leqslant C_0^{-\tau(B,u)/2}.$$

We are now in position to apply the interpolation formula of Chapter VII, § 5, to the function of one variable

$$g(z) = \sum a^{(J)}(\lambda^{(J)})\Psi_{\lambda, m'}\left(\frac{z}{2^J}\right) e^{u_1 z\lambda_1} \dots e^{u_{r-1} z\lambda_{r-1}} e^{(\beta_1 u_1 + \dots + \beta_{r-1} u_{r-1}) z\lambda_r}$$

$$= G_{m'}(z, u_1 z, \dots, u_{r-1} z) .$$

We are trying to show that

$$G_{m'}\left(\frac{n'}{2}, \frac{n' u_1}{2}, \dots, \frac{n' u_{r-1}}{2}; \frac{n' u_r}{2}\right) = 0$$

for all integers n' with $1 \leqslant n' \leqslant 2^{J+1}N$. We therefore estimate $g(n'/2)$. The radii R_1 and R_2 are of the same order of magnitude, namely $C2^J N$, and we choose a large radius R, say

$$R = M_* 2^J N .$$

We find:

$$\left|g\left(\frac{n'}{2}\right)\right| \leqslant \frac{\|g\|_R C^{MN}}{C^{MN \operatorname{Log} M_*}} + C^{MN \operatorname{Log} M_*} \max |D^K g(n)|$$

where the maximum of the derivatives is taken for $0 \leqslant \kappa \leqslant M/2^{J+1}$ and for odd n, $1 \leqslant n \leqslant 2^J N$.

Expressing $D^\kappa g(n)$ in terms of $D^{(k)} G_{m'}$, we see that the second term on the right is small. As for the first term, we have the easy estimate

$$\|g\|_R \leqslant C^{R \sum U_j L_j / 2^J} \leqslant C^{MN} .$$

We may therefore conclude that there is a constant C depending on the usual r, $[K: \mathbf{Q}]$, such that for M_* large (and $C_0 = \exp(M_*^{r+2})$)

$$|g(n'/2)| \leqslant M_*^{-CMN} .$$

By Lemma 4.1, the algebraic number

$$\xi_{m', n'} = G_{m'}(n'/2, n' u_1/2, \dots, n' u_{r-1}/2; n' u_r/2)$$

is within $C_0^{-\tau(B, u)/2}$ of $g(n'/2)$, and consequently this algebraic number $\xi_{m', n'}$ satisfies same sort of upper estimate. If it is not zero, we also get a lower bound for its absolute value in terms of its size, namely

$$C^{-MN} \leqslant |\xi_{m', n'}| \leqslant C^{-MN \operatorname{Log} M_*} .$$

This is a contradiction, which shows that $\xi_{m',n'} = 0$.

We use these equations only for odd n', in the usual range $1 \leqslant n' \leqslant 2^{J+1}N$. Using our assumption

$$[K(\alpha_1^{1/2}, \ldots, \alpha_r^{1/2}) : K] = 2^r ,$$

we see that the equation $\xi_{m',n'} = 0$ splits into 2^r equations. We express $\xi_{m',n'}$ in terms of the basis

$$\alpha_1^{n'l_1/2} \ldots \alpha_r^{n'l_r/2}, \quad l_j = 0, 1 ,$$

of the field $K(\alpha_1^{1/2}, \ldots, \alpha_r^{1/2})$ over K:

$$\xi_{m',n'} = \sum_{l_1=0}^{1} \cdots \sum_{l_r=0}^{1} \xi_{l,m',n'} \alpha_1^{n'l_1/2} \ldots \alpha_r^{n'l_r/2} .$$

Thus $\xi_{l,m',n'} \alpha_1^{n'l_1/2} \ldots \alpha_r^{n'l_r/2}$ is obtained from $\xi_{m',n'}$ by considering in the sum $\sum_{(\lambda^{(J)})}$ (of the definition of $G_{m'}$) only those $(\lambda^{(J)})$ for which $\lambda_j \equiv l_j \bmod 2$, $(1 \leqslant j \leqslant r)$. We get the equations

$$\xi_{l,m',n'} = 0 ,$$

with

$$\xi_{l,m',n'} =$$

$$\sum_{\substack{0 \leqslant \lambda_{-1} \leqslant L_{-1} \\ 0 \leqslant \lambda_0 \leqslant L_0}} \sum_{\substack{0 \leqslant \lambda_j \leqslant L_j/2^J \\ \lambda_j \equiv l_j \bmod 2 \\ 1 \leqslant j \leqslant r}} a^{(J)}(\lambda^{(J)}) \Psi(n'/2^{J+1}, m', \lambda^{(J)}, r-1) \cdot \alpha_1^{n'(\lambda_1 - l_1)/2} \ldots \alpha_r^{n'(\lambda_r - l_r)/2} .$$

We choose $l_1, \ldots, l_n = 0, 1$ in such a way that at least one of the numbers

$$a^{(J)}(\lambda^{(J)}), \qquad \lambda_j \equiv l_j \bmod 2 \quad \text{for } 1 \leqslant j \leqslant r$$

is non-zero. We denote by $a^{(J+1)}(\mu)$ the numbers thus obtained with

$$\mu = (\lambda_{-1}, \lambda_0, \mu_1, \ldots, \mu_r),$$

$$\lambda_j = l_j + 2\mu_j, \quad (1 \leqslant j \leqslant r) .$$

Therefore we have the equations

$$\sum_{(\mu)} a^{(J+1)}(\mu) \frac{1}{m_0'!} D_0^{m_0'} \Delta^{\lambda_0} \left(\frac{n'}{2^{J+1}} + \lambda_{-1}, L_{-1} \right) \prod_{j=1}^{r-1} \Delta(y_j(\mu), m_j') \alpha_1^{\mu_1 n'} \cdots \alpha_r^{\mu_r n'} = 0 ,$$

where

$$y_j(\mu) = l_j + 2\mu_j + \beta_j(l_r + 2\mu_r)$$
$$= 2(\mu_j + \mu_r\beta_j) + l_j + 2l_r\beta_j .$$

As $y_j(\mu)$ is a polynomial of degree 1 in $(\mu_j + \mu_r\beta_j)$, from the remark made in the last section we deduce that the same equations hold with $y_j(\mu)$ replaced by

$$x_j(\mu) = \mu_j + \mu_r\beta .$$

We conclude the proof of Theorem 1.2 by writing $(\lambda^{(J+1)}) = (\mu)$.

We end this section by some comments on the choice of the parameters. The size inequalities involve essentially

$$\exp\{C(M\log B + JL_0L_{-1} + ML_{-1} + \sum L_jNU_j)\} .$$

It turns out that the other terms like L_0N and $L_r\log B$ are smaller.

In the interpolation formula, our choice was $R = M_*2^JN$; then the dominating term for $\|g\|_R$ is $R\sum L_jU_j/2^J$. This term has to be smaller than $CMN\log(R/CN2^J)$.

Therefore we choose N, L, M in such a way that

$$\max\{M\log B, L_0L_{-1}\log U_{r-1}, ML_{-1}, M_*N\sum L_jU_j\}$$

is small compared with

$$MN\log M_* .$$

The cheapest way of achieving such an aim (apart from M_*, which is irrelevant) is to define

$$N = L_{-1} = [\log B]$$

$$L_0 = M/\log U_{r-1}$$

$$L_j = M/M_*U_j, \quad (1 \leqslant j \leqslant r) .$$

We now write the assumption which is needed to apply Siegel's lemma: the number $L_{-1}L_0L_1\dots L_r$ must be larger than NM^r. This yields

$$M \geqslant M_*^{r+1}U_1\dots U_r\log U_{r-1} .$$

Finally, for Lemma 4.1 and for the second term in the interpolation formula, we have to assume that

$$|\beta_1u_1 + \dots + \beta_{r-1}u_{r-1} - u_r|$$

is small compared with $MN \log M_*$. This last condition settles our choice

$$C_0^{-\tau(B,u)} = \exp\{-M_*^{r+2}(\log B)U_1 \dots U_r \log U_{r-1}\}.$$

Obviously, our proof of Theorem 1.2 extends to the case of any prime p in place of 2, provided that we allow the constant C_0 to depend on p.

Up to now we used the rationality of the β's only once, namely in the estimates for the denominator for the Feldman polynomials. (These estimates occur at two places in the proof, first for Siegel's lemma, then in the lower bound for $|\xi_{m',n'}|$.)

§ 5. Final Descent

In this section we perform the reduction which will enable us to derive Theorem 1.1 from Theorem 1.2.

Let K be a number field. Let γ_0 be a generator of the group of roots of unity of K, and $\gamma_1, \dots, \gamma_r$ multiplicatively independent elements of K^*. Let Γ be the multiplicative group generated by $\gamma_0, \dots, \gamma_r$, and let Γ' be the division group of Γ in K^*. We let $\gamma_0' = \gamma_0$.

For $j = 0, \dots r$ we let $U_j \geqslant e$ be an upper bound for $\log H_K(\gamma_j)$. We assume

$$U_1 \leqslant U_2 \leqslant \dots \leqslant U_r.$$

We repeat as a lemma the corollary of Theorem 5.2 of Chapter IV.

Lemma 5.1. *There exist $\gamma_1', \dots, \gamma_r'$ free generators of Γ' modulo the torsion subgroup (γ_0), such that*

$$\log H_K(\gamma_j') \leqslant U_1 + \dots + U_j \text{ for } j = 1, \dots, r, \tag{1}$$

and there exist rational integers $m_{j,k}$ $(j = 1, \dots, r;\ k = -1, \dots, j)$ satisfying

$$m_{j,-1} > 0, \quad \textit{and} \quad \max_k m_{j,k} \leqslant C_1 U_j^{j+1} \quad \textit{for } j = 1, \dots, r$$

such that

$$m_{j,-1} \log \gamma_j = \sum_{k=0}^{j} m_{j,k} \log \gamma_k'. \tag{2}$$

The logarithms are taken as principal valued, and C_1 is effectively computable, depending only on r and $[K:\mathbf{Q}]$.

We observe that in the proof of this result, we constructed $\gamma_1', \dots, \gamma_j'$, independently of $\gamma_{j+1}, \dots, \gamma_r$. The same goes for the coefficients $m_{j,k}$.

Lemma 5.2. *Let p be a prime such that K contains the p-th roots of unity. There exist* $\gamma_1^0, \ldots, \gamma_r^0$ *in* Γ', *and there exist rational integers*

$$m_{j,k}^0 \; (j = 1, \ldots, r; k = -1, \ldots, j)$$

satisfying:

$$[K(\gamma_0^{1/p}, (\gamma_1^0)^{1/p}, \ldots, (\gamma_r^0)^{1/p}) : K] = p^{r+1} \tag{3}$$

$$\log H_K(\gamma_j^0) \leqslant U_1 + \cdots + U_j \quad (j = 1, \ldots, r) \tag{4}$$

$$m_{j,-1}^0 \log \gamma_j = \sum_{k=0}^{j} m_{j,k}^0 \log \gamma_k^0 \quad (j = 1, \ldots, r \quad \text{and} \quad \gamma_0^0 = \gamma_0) \tag{5}$$

$$m_{j,-1}^0 > 0 \quad \text{and} \; \max_{-1 \leqslant k \leqslant j} |m_{j,k}^0| \leqslant C_1 U_j^{j+C_2+1} \quad (j = 1, \ldots, r) \tag{6}$$

$$\max_{-1 \leqslant k \leqslant r} |m_{r,k}^0| \leqslant \left(\frac{rU_r}{\log H_K(\gamma_r^0)}\right)^{C_2} C_3 U_{r-1}^{r+1}. \tag{7}$$

Here C_2, C_3 *are effectively computable constants depending only on r, p and* $[K : \mathbf{Q}]$.

(We shall prove this result with $C_2 = \dfrac{\log(3rpD^2)}{\log 3/2}$ and $C_3 = (6rC_1)^{r+1}$.)

Proof. We first use Lemma 5.1 for the subgroup Γ_{r-1} of K^* generated by $\gamma_1, \ldots, \gamma_{r-1}$. Let $\gamma_1', \ldots, \gamma_{r-1}'$ be free generators of Γ_{r-1}' as constructed in the proof of Lemma 1. Thus we have

$$m_{j,-1} \log \gamma_j = \sum_{k=0}^{j} m_{j,k} \log \gamma_k' \quad j = 1, \ldots, r-1,$$

where the integers $m_{j,k}$ ($1 \leqslant j \leqslant r-1$ and $0 \leqslant k \leqslant j$) satisfy (6).

We shall now prove:

Let V be a real number, with $U_{r-1} \leqslant V \leqslant U_r$. *Let* η *be an element of* Γ' *which is multiplicatively independent of* $\gamma_1, \ldots, \gamma_{r-1}$, *with* $\log H_K(\eta) \leqslant V$. *Then there exists* $\eta' \in \Gamma'$ *and* $m_{-1}, m_0, \ldots, m_r$ *in* $\mathbf{Z}$ *such that:*

$$[K(\gamma_0^{1/p}, \gamma_1'^{1/p}, \ldots, \gamma_{r-1}'^{1/p}, \eta'^{1/p}) : K] = p^{r+1} \tag{8}$$

$$\log H_K(\eta') \leqslant U_1 + \cdots + U_{r-1} + V \tag{9}$$

$$m_{-1} \log \eta = \sum_{k=0}^{r-1} m_k \log \gamma_k' + m_r \log \eta' \tag{10}$$

(11) $\quad m_{-1} > 0 \quad and \quad \max_{-1 \leqslant k \leqslant r} |m_k| \leqslant C_1 V^{C_4} \quad with \quad C_4 = \max(r+1, C_2)$

(12) $$\max_{-1 \leqslant k \leqslant r} |m_k| \leqslant \left(\frac{rV}{\log H_K(\eta')}\right)^{C_2} C_3 U_{r-1}^{r+1} .$$

For $V = U_r$ this will give the desired result. On the other hand, for

$$U_{r-1} \leqslant V \leqslant 6r^2 U_{r-1} ,$$

this statement follows from Lemma 5.1 and Kummer theory.

We shall prove the preceding statement by induction on the integral part of V. From now on, we assume that the statement holds for

$$6r^2 U_{r-1} \leqslant V \leqslant V_0$$

with some integer $V_0 \leqslant U_r - 1$, and we prove it for $V_0 < V \leqslant V_0 + 1$.

Thus let $\eta' \in \Gamma'$ be multiplicatively independent of $\gamma_1, \ldots, \gamma_{r-1}$, with

$$\log H_K(\eta) = V_0 + 1 .$$

If (8) holds with $\eta' = \eta$, then we choose $m_{-1} = m_r = 1$, and

$$m_0 = \cdots = m_{r+1} = 0 .$$

Now assume

$$[K(\gamma_0^{1/p}, \gamma_1'^{1/p}, \ldots, \gamma_{r-1}'^{1/p}, \eta^{1/p}) : K] \neq p^{r+1} .$$

By Kummer theory, there exists an element $\eta_1 \in K^*$ and integers $t_1, \ldots, t_{r-1}$ between 0 and $p-1$ such that

(13) $$\eta = \gamma_1'^{t_1} \ldots \gamma_{r-1}'^{t_{r-1}} \eta_1^p .$$

Therefore $\eta_1 \in \Gamma'$, and from the properties of H_K we get

$$p \log H_K(\eta_1) = \log H_K(\eta_1^p) \leqslant \log H_K(\eta) + \sum_{j=1}^{r-1} t_s \log H_K(\gamma_j') .$$

It follows that

$$\log H_K(\eta_1) = \frac{1}{p} V + (r-1)^2 U_{r-1} \leqslant \tfrac{2}{3} V .$$

By the induction hypothesis there exist $\eta_1' \in \Gamma'$ and

$$m'_{-1}, m'_0, \ldots, m'_r \in \mathbf{Z}$$

such that the corresponding properties (8), (9), (10), (11), (12) hold. In particular,

$$\log H_K(\eta'_1) \leqslant U_1 + \cdots + U_{r-1} + V$$

$$m'_{-1} \log \eta_1 = \sum_{k=0}^{r-1} m'_k \log \gamma'_k + m'_r \log \eta'_1$$

$$\max_{-1 \leqslant k \leqslant r} |m'_k| \leqslant C_1 (2/3)^{C_2} V^{C_4}$$

$$\max_{-1 \leqslant k \leqslant r} |m'_k| \leqslant \left(\frac{2}{3}\right)^{C_2} \left(\frac{rV}{\log H_K(\eta'_1)}\right)^{C_2} C_3 U^r_{r-1} .$$

We now come back to (13). There exists an integer t_0 such that

$$\log \eta = t_0 \log \gamma_0 + \sum_{j=1}^{r-1} t_s \log \gamma'_s + p \log \eta_1 ,$$

and we easily deduce as in the proof of Lemma 5.1 that

$$|j_0| \leqslant D^2(rp + 1) .$$

Now we obtain

$$m_{-1} \log \eta = \sum_{j=0}^{r-1} m_j \log \gamma'_j + m_r \log \eta'_1 ,$$

with

$$m_{-1} = m'_{-1}, \qquad m_r = pm'_r, \quad \text{and} \quad m_j = t_j m'_{-1} + pm'_j \quad (0 \leqslant j \leqslant r-1) .$$

Therefore

$$\max_{-1 \leqslant j \leqslant r} |m_j| \leqslant \left(\max_{-1 \leqslant j \leqslant r} |m'_j| \right)(3rpD2) .$$

We now remark that

$$3rpD^2 \leqslant (\tfrac{3}{2})^{C_2} .$$

The desired result follows.

We shall use only a weak consequence of (6) and (7), namely

(14) $$(\log \max_{j,k} |m^0_{j,k}|)(\log H_K(\gamma^0_r) + 1) \leqslant C_5 U_r \log U_{r-1} ,$$

with, say, $C_5 = 1 + r(r+1)C_2 C_3$.

Lemma 5.3. *Let*

$$\Lambda = \beta_1 u_1 + \cdots + \beta_n u_n \neq 0$$

be a linear form in n logarithms with rational coefficients. Let

$$B = H_{\mathbf{Q}}(1, \beta_1, \ldots, \beta_n).$$

Suppose that

$$U_1 \leqslant \cdots \leqslant U_n \quad and \quad U_{n-1} \leqslant B.$$

Let

$$\tau(B, u) = (\log B) U_1 \ldots U_n \log U_{n-1}.$$

There exists a linear form

$$\Lambda^0 = \beta_0^0 u_0^0 + \cdots + \beta_{r-1}^0 u_{r-1}^0 - u_r^0 \neq 0$$

with

$$\exp u_j^0 = \alpha_j^0 \in K, \quad \beta_j^0 \in \mathbf{Q}, \quad r \leqslant n,$$

such that

$$[K((\alpha_0^0)^{\frac{1}{2}}, \ldots, (\alpha_r^0)^{\frac{1}{2}}) : K] = 2^{r+1}$$

and such that

$$\frac{\log |\Lambda|}{\tau(B, u)} \geqslant C \frac{\log |\Lambda^0|}{\tau(B^0, u^0)},$$

where C is an effectively computable constant depending only on n and $[K : \mathbf{Q}]$.

We have used the obvious notation:

$$B^0 = H_{\mathbf{Q}}(1, \beta_0^0, \ldots, \beta_{r-1}^0), \qquad U_j^0 \geqslant \max\{\log H_K(\alpha_j^0), e\}$$

and U_j^0 are such that

$$U_0^0 \leqslant U_1^0 \leqslant \cdots \leqslant U_r^0.$$

Finally,

$$\tau(B^0, u^0) = (\log B^0) U_0^0 \ldots U_r^0 (\log U_{r-1}^0).$$

Proof of Lemma 5.3. We first define $u_0 = \log \gamma_0$, where γ_0 is a root of unity of maximal order in K. Then we select inductively

$$u_{j0} = u_0, u_{j1}, \ldots, u_{j_s} \quad \mathbf{Q}\text{-linearly independent}$$

such that for $j_k < j < k_{k+1}$, the element u_j is a linear combination of $u_{j0}, \ldots, u_{j_k}$ with rational coefficients.

Substituting, we get

$$\Lambda = \sum_{\nu=0}^{s} \beta'_\nu u_{j_\nu} .$$

Define

$$B' = H_{\mathbf{Q}}(1, \beta'_0, \ldots, \beta'_s) .$$

We remark that:

Either $j_s = n$, in which case $\beta'_s = \beta_n$, and (since $U_{n-1} \leqslant B$)

$$B' \ll BU^s_{n-1} \log\log U_{n-1}, \quad \text{whence} \quad \log B' \ll \log B ;$$

or $j_s < n$, in which case we need only the weak bound

$$\log B' \ll U_n \log B .$$

In both cases, we get, putting $u' = (u_{j0}, \ldots, u_{j_s})$:

$$\begin{aligned}\tau(B', u') &= (\log B')U_{j0} \cdots U_{j_s} \log U_{j_s - 1} \\ &\ll (\log B)U_1 \cdots U_n \log U_{n-1} = \tau(B, u) .\end{aligned}$$

We now use Lemma 5.2 for the subgroup Γ of K^* generated by the numbers $\exp u_{jk}$ with $0 \leqslant k \leqslant s$. We can find

$$\alpha^0_0 = \gamma_0, \alpha^0_1, \ldots, \alpha^0_s$$

such that

$$\Lambda = \sum_{\nu=0}^{s} \beta'_\nu \sum_{k=0}^{\nu} \frac{m^0_{\nu,k}}{m^0_{\nu,-1}} \log \alpha^0_k = \sum_{k=0}^{s} \beta''_k \log \alpha^0_k$$

with

$$\beta''_k = \sum_{\nu=k}^{s} \beta'_\nu \frac{m^0_{\nu,k}}{m^0_{\nu,-1}} .$$

We define

$$r = \max \text{ of those } k \text{ such that } \beta_k'' \neq 0\,,$$

and

$$\Delta^0 = -\Lambda/\beta_r'', \quad \beta_k^0 = -\beta_k''/\beta_r'' \quad \text{for} \quad 0 \leqslant k \leqslant r, \quad u_k^0 = \log \alpha_k^0\,.$$

Since

$$B^0 = H_{\mathbf{Q}}(1, \beta_0^0, \ldots, \beta_{r-1}^0) \leqslant H_{\mathbf{Q}}(1, \beta_0'', \ldots, \beta_{r-1}'', \beta_r'') \ll B \max_{\substack{-1 \leqslant k \leqslant v \\ 0 \leqslant v \leqslant s}} |m_{v,k}^0|$$

we have by (14)

$$(\log B^0) U_{js}^0 \ll (\log B) U_n + U_n \log U_{n-1} \ll (\log B) U_n\,,$$

and consequently

$$\tau(B^0, u^0) \ll \tau(B, u)\,.$$

Lemma 5.3 follows immediately.

Proof of Theorem 1.1. We assume, as we may without loss of generality, that we have a linear form as in Lemma 5.3, with

$$n \geqslant 2 \quad \text{and} \quad U_1 \leqslant \cdots \leqslant U_n\,.$$

If $B \leqslant U_{n-1}$ then the Liouville inequality yields

$$|\Lambda| \geqslant C^{-U_{n-1} U_n}$$

which implies the theorem. If $B \geqslant U_{n-1}$ then the desired inequality follows from Theorem 1.2 and Lemma 5.3.

Bibliography

[A] W. ADAMS, Transcendental numbers in the p-adic domain, Amer. J. Math. **88** (1966) pp. 279–308

[Bach] M. BACHMAKOV, Un théoreme de finitude sur la cohomologie des courbes élliptiques, C.R. Acad. Sci. Paris **270** (1970) pp. 999–1001

[Ba 1] A. BAKER, Transcendental number theory, Cambridge University Press, 1975

[Ba 2] ——, A sharpening of the bounds for linear forms in logarithms, Acta Arithmetica **XXI** (1972) pp. 117–129

[Ba 3] ——, A sharpening of the bounds for linear forms in logarithms III, Acta Arith. **XXVII** (1975) pp. 247–252

[Ba 4] ——, Contributions to the theory of diophantine equations, I and II, Phil. Trans. Royal Soc. London Series A. Math and Phys. Sci. No. 1139 Vol. 263 pp. 173–208

[Ba 5] ——, On the periods of the Weierstrass $\wp$-function, Proc. Rome Confer. on number theory, 1968, Academic Press, London, 1970, pp. 155–174

[Ba 6] ——, On the quasi-periods of the Weierstrass ζ-function, Göttinger Nachr. (1969) No. 16, pp. 145–157

[Ba 7] ——, The theory of linear forms in logarithms, Transcendence theory and applications, Proceedings of Cambridge Conference, 1976, Academic Press, 1977

[BC] A. BAKER and J. COATES, Integer points on curves of genus 1, Proc. Cambridge. Philos. Soc. **67** (1970) pp. 595–602

[BS] A. BAKER and H. STARK, On a fundamental inequality in number theory, Ann. of Math. **94** (1971) pp. 190–199

[Be 1] D. BERTRAND, Algebraic values of p-adic elliptic functions; Advances in transcendence theory, ed. A. Baker, D. Masser, Academic Press, 1977 (chapter 9)

[Be 2] ——, Séries d'Eisenstein et transcendance, Bull. Soc. Math. France **104**, 1976, pp. 309–321

[Be 3] ——, Sous-groupes à un paramètre p-adique de variétés de groupe, Inventiones math. **40**, 1977, pp. 171–193

[Be 4] ——, Approximations diophantiennes p-adiques sur les courbes elliptiques admettant une multiplication complexe, Compositio math., to appear

[Be 5] ——, Transcendance et lois de groupes algébriques, Séminaire Delange-Pisot-Poitou, 18e année, 76/77, n^0 1 (Secrétariat mathématique, Paris)

[B-M] P. Blanksby and H. Montgomery, Algebraic integers near the unit circle, Acta Arith. **18** (1971) pp. 355–369

[Ca 1] J. W. S. Cassels, Diophantine equations with special reference to elliptic curves, J. London Math. Soc. **41** (1966) pp. 193–291

[Ca 2] ——, A note on division values of $\wp(u)$, Proc. Cambridge Philos. Soc. **45** Pt. 2 (1949) pp. 167–172

[Ch] C. Chabauty, Démonstration de quelques lemmes de rehaussement, C. R. Acad. Sci. Paris **217** (1943) pp. 413–415

[ChW] C. Chevalley and A. Weil, Un théorème d'arithmétique sur les courbes algébriques, C.R. Acad. Sci. Paris (1930) pp. 570–572

[C-T] P. Cijsouw and R. Tijdeman, An auxiliary result in the theory of transcendental numbers II, Duke Math. J. **42** (1975) pp. 239–247

[C-W] P. Cijsouw and M. Waldschmidt, Linear forms and simultaneous approximations, Compositio Math. **34** (1977) pp. 173–197

[Co 1] J. Coates, An effective p-adic analogue of a theorem of Thue, Acta Arith.: Part I, **XV** (1969) pp. 279–305; Part II, **XVI** (1970) pp. 399–412; Part III, **XVI** (1970) pp. 425–435

[Co 2] ——, Construction of rational functions on a curve, Proc. Cambridge Philos. Soc. **68** (1970) pp. 105–602

[Co 3] ——, Linear forms in the periods of the exponential and elliptic functions, Invent. Math. **12** (1971) pp. 290–299

[Co 4] ——, Linear relations between $2\pi i$ and the periods of two elliptic curves, Diophantine approximation and its applications, Academic Press, London (1973) pp. 77–99

[Co 5] ——, The transcendence of linear forms in $\omega_1, \omega_2, \eta_1, \eta_2, 2\pi i$, Amer. J. Math. **93** (1971) pp. 385–397

[Co 6] ——, An application of the division theory of elliptic functions to diophantine approximation, Invent. Math. **11** (1970) pp. 167–182

[CL] J. Coates and S. Lang, Diophantine approximation on abelian varieties with complex multiplication, Invent. (1976) pp. 129–133

[C-T] J. L. Colliot-Thélène, Variation de la hauteur sur une famille de courbes élliptiques particulière, Acta Arith. **XXXI** (1976) pp. 1–16

[Da] H. Davenport, Analytic methods for diophantine equations and diophantine inequalities, Lecture Notes, Univ. Michigan 1962

[De 1] V. A. Demjanenko, On torsion points of elliptic curves, Math. USSR-Izv. **4** (1970) No. 4, pp. 765–783

[De 2] ——, Torsion of elliptic curves, Math. USSR-Izv., **5** (1971) No. 2, pp. 289–318

[De 3] ——, On the uniform boundedness of the torsion on elliptic curves over algebraic number fields, Math. USSR-Izv. **6** (1972) No. 3, pp. 477–490

[De 4] ——, Rational points of a class of algebraic curves, Izv. Akad. Nauk SSSR Ser. Mat. **28** (1964) pp. 1363–1390, English translation (2) **59** (1966) pp. 82–110

[De 5] ——, On the representation of numbers by binary quadratic forms, Math. USSR-Sb. Vol. 9 (1969) No. 3 pp. 415–522

[De 6] ——, On Tate height and the representation of numbers by binary forms, Math. USSR-Izv. **8** (1974) No. 3 pp. 463–476

[De 7] ——, Estimate of the remainder term in Tate's formula, Mat. Zametki **3** (1968) pp. 271–278, translated in Math. Notes **3** (1968) pp. 173–177

[De 8] ——, On Mordell's conjecture, Math. USSR-Izv. **8** (1974) No. 6, pp. 1181–1189

[Fe 1] N. I. Feldman, The approximation of certain transcendental numbers I and II, Izv. Akad. Nauk SSSR Ser. Mat. **15** (1951) pp. 53–74 and 153–176 (AMS Translation 1966)

[Fe 2] ——, Arithmetic properties of values of elliptic functions, Izv. Akad. Nauk SSSR Ser. Mat. **22** (1958) pp. 563–576 (AMS Translation 1966)

[Fe 3] ——, On the measure of transcendence of π, Izv. Akad. Nauk SSSR **24** (1960) pp. 357–368 (AMS Translation 1966)

[Fe 4] ——, Improved estimate for a linear form of the logarithm of algebraic numbers, Mat. Sb. **77** (1968) pp. 423–436, AMS Translations pp. 393–406

[Fe 5] ——, An elliptic analogue of an inequality of Gelfond, Trudy Moskov. Mat. Obšč. **18** (1968) pp. 65–76, AMS Transl. (1968) pp. 71–83

[Fe 6] ——, An effective refinement of the exponent in Liouville's theorem, Izv. Akad. Nauk **35** (1971) pp. 973–990, AMS Transl. **5** (1971) pp. 985–1002

[Fo] J. M. Fontaine, Groupes finis commutatifs sur les vecteurs de Witt, C.R. Acad. Sci. Paris **280** (1975) pp. 1423–1425

[Ge] A. O. Gelfond, Transcendental and algebraic numbers, Translation, Dover Publications, 1960 (Russian edition, 1951)

[Ha] H. Hasse, Rein arithmetischer Beweis des Siegelschen Endlichkeitssatzes für binare diophantische Gleichungen im Spezialfall des Geschlechts 1, Abh. D. Akad. Wiss. Berlin (1951) No. 2

[Ig] J. Igusa, Analytic groups over complete fields, Proc. Nat. Acad. Sci. USA, Vol. 42, No. 8 (1956) pp. 540–541

[K] D. Kubert, Quadratic relations for generators of units in the modular function field, Math. Ann. **225** (1977) pp. 1–20

[KL] D. Kubert and S. Lang, Units in the modular function field, Math. Ann. (1975) Part I, pp. 67–96, and II pp. 175–189

[L 1] S. Lang, Diophantine Geometry, Wiley-Interscience, 1962

[L 2] ——, Elliptic functions, Addison Wesley, 1973

[L 3] ——, Introduction to transcendental numbers, Addison Wesley, 1966

[L 4] ——, Introduction to diophantine approximations, Addison Wesley, 1966

[L 5] ——, Diophantine approximation on toruses, Amer. J. Math. **86** (1964) pp. 521–533

[L 6] ——, On a theorem of Mahler, Mathematika **7** (1960) pp. 139–140

[L 7] ——, Integral points on curves, Publ. IHES (1960) pp. 27–43

[L 8] ——, Division points on curves, Ann. Mat. Pura Appl. IV, Tomo LXX (1965) pp. 229–234

[L 9] ——, Les formes bilinéaires de Néron et Tate, Sem. Bourbaki 274, 1963–1964

[L 10] ——, Division points of elliptic curves and abelian functions over number fields, Amer. J. Math. **97** No. 1 (1972) pp. 124–132

[L 11] ——, Diophantine approximation on abelian varieties with complex multiplication, Advances in Math. **17** No. 3 (1975) pp. 281–336

[L 12] ——, Sur la conjecture de Catalan, d'après Tijdeman, Seminaire Delange-Poitou-Pisot, Paris, 1976

[L–T] S. Lang and H. Trotter, Primitive points on elliptic curves, Bull. AMS **83** (1977) pp. 289–292

[Li 1] P. Liardet, Sur une conjecture de Serge Lang, C.R. Acad. Sci. Paris **279** (1974) pp. 435–437

[Li 2] ——, Sur une conjecture de Serge Lang, Soc. Math. France, Astérisque **24–25** (1975) pp. 187–210

[Lub 1] J. Lubin, One parameter formal Lie groups over p-adic integer rings, Ann. of Math. **80** (1964) pp. 464–484

[Lub 2] ——, Finite subgroups and isogenies of one parameter formal Lie groups, Ann. of Math. **85** (1967) pp. 296–302

[Lut] E. Lutz, Sur l'équation $y^2 = x^3 - Ax - B$ dans les corps p-adiques, J. Reine Angew. Math. **177** (1937) pp. 237–247

[Mah 1] K. Mahler, Ein Analogon zu einem Schneiderschen Satz, Proc. Kon. Akad. Wetensch. Amsterdam **39** (1936) pp. 633–640

[Mah 2] ——, Über transzendente p-adische Zahlen, Comp. Math. **2** (1935) pp. 259–275

[Mah 3] ——, On some inequalities for polynomials in several variables, J. London Math. Soc. **37** (1962) pp. 341–344

[Mah 4] ——, Über die rationalen Punkte auf Kurven vom Geschlecht Eins, J. Reine Angew. Math. **170** (1934) pp. 168–178

[Man 1] J. MANIN, The refined structure of the Néron-Tate height, Mat. Sb. (1970), AMS Transl. pp. 325–342

[Man 2] ——, The Tate height of points on an abelian variety, its variants and applications, Izv. Akad. Nauk SSSR Ser. Mat. **28** (1964), AMS Transl. (2) **59** (1966) pp. 82–110

[Man 3] ——, The p-torsion of elliptic curves is uniformly bounded, Izv. Akad. Nauk SSSR (1969) AMS Transl. pp. 433–438

[Man 4] ——, Cyclotomic fields and modular curves, Russian Math. Surveys Vol. 26 No. 6 (1971)

[Mas 1] D. MASSER, Elliptic functions and transcendence, Springer Lecture Notes **437** (1975)

[Mas 2] ——, Linear Forms in algebraic points of abelian functions, I, Math. Proc. Cambridge Philos. Soc. **77** (1975) 499–512; II, Math. Proc. Cambridge Philos. Soc. **79** (1976) 55–70; III, J. London Math. Soc. (1976) 549–564

[Mat] A. MATTUCK, Abelian varieties over p-adic ground fields, Ann. of Math. **62** (1955) pp. 92–119

[Maz] B. MAZUR, Modular curves and the Eisenstein ideal, Publ. IHES (1977)

[Mo] L. J. MORDELL, Diophantine Equations, Academic Press, 1969

[Ne 1] A. NERON, Modèles minimaux des variétés abéliennes sur les corps locaux et globaux, Publ. IHES **21** (1964)

[Ne 2] ——, Quasi-fonctions et hauteurs sur les variétés abéliennes, Ann. of Math. **82** (1965) pp. 249–331

[OT] F. OORT and J. TATE, Group schemes of prime order, Ann. Sci. ENS, série 4, § 3, (1970) pp. 1–21

[Ra] M. RAYNAUD, Schémas en groupes de type $(p, \ldots, p)$, Bull. Soc. Math. France **102** fasc. 3 (1974) pp. 241–280

[Ri] K. RIBET, Dividing rational points on abelian varieties of CM type, Compositio Math. **33** (1976) pp. 69–74

[Ro] P. ROBBA, Lemme de Schwarz p-adique pour plusieurs variables,

[Sa] H. SAH, Automorphisms of finite groups, J. Algebra **10** (1968) pp. 47–68, especially p. 60, Proposition 2.7

[Scha] S. SCHANUEL, Heights in number fields, to appear Bull. Soc. Math. France

[Sch 1] T. SCHNEIDER, Zur Theorie der Abelschen Funktionen und Integrale, J. Reine Angew. Math. (1941) pp. 110–128

[Sch 2] ——, Ein Satz über ganzwertige Funktionen als Prinzip für Transzendenzbeweise, Math. Ann. **121** (1949) pp. 131–140

[Se 1] J. P. SERRE, Dépendance d'exponentielles p-adiques, Séminaire Delange-Pisot-Poitou (1965–1966)

[Se 2] ——, Proprietes galoisiènnes des points d'ordre fini des courbes élliptiques, Invent. Math. **15** (1972) pp. 259–331

[Se 3] ——, Lie algebras and Lie groups, Benjamin, 1964

[Si 1] C. L. SIEGEL, Über einige Anwendungen diophantischer Approximationen, Abh. Preuss. Akad. Wiss. (1929) pp. 1–41

[Si 2] ——, Abschätzung von Einheiten, Nachr. Akad. Wiss. Göttingen (1969) pp. 71–86

[Si 3] ——, The integer solutions of the equation $y^2 = ax^n + bx^{n-1} + \cdots + k$, J. London Math. Soc. **1** (1926) pp. 66–68

[Si 4] ——, Die Gleichung $ax^n - by^n = c$, Math. Ann. **114** (1937) pp. 57–68

[St 1] H. STARK, A transcendence theorem for class number problems I and II, Ann. Math. **94** (1971) pp. 153–173 and **96** (1972), pp. 174–209

[St 2] ——, Effective estimates of solutions of some diophantine equations, Acta Arith. **24** (1973) pp. 251–259

[St 3] ——, Further advances in the theory of linear forms in logarithms, Diophantine approximation and its applications, Academic Press, London (1973) pp. 255–293

[Ta 1] J. TATE, The arithmetic of elliptic curves, Invent. Math. **23** (1974) pp. 179–206

[Ta 2] ——, Algorithm for determining the type of a singular fiber in an elliptic pencil, Modular Functions in one variable IV, Springer Lecture Notes **476** (Antwerp Conference)

[Ti 1] R. TIJDEMAN, On the equation of Catalan, Acta Arith. **XXIX** (1976) pp. 197–209

[Ti 2] ——, An auxiliary result in the theory of transcendental numbers, J. Number Theory **5** (1973) pp. 80–94

[VdP 1] A. J. Van der POORTEN, On Baker's inequality for linear forms in logarithms, Math. Proc. Cambridge Philos. Soc. **80** (1976) pp. 233–248

[VdP 2] ——, Linear forms in logarithms in the p-adic case, Transcendence theory and its applications, Academic Press, 1977

[VdP 3] ——, Effectively computable bound for the solution of certain diophantine equations, Acta Arith., to appear

[VdP–L 1] A. J. Van der Poorten and J. H. Loxton, Computing the effectively computable bound in Baker's inequality for linear forms in logarithms, Bull. Austral. Math. Soc. **15** (1976) pp. 33–57

[VdP–L 2] ——, Multiplicative relations in number fields, Bull. Austral. Math. Soc. **16** (1977) pp. 83–98

[Wa 1] M. Waldschmidt, Nombres transcendents, Springer Lecture Notes **402** (1974)

[Wa 2] ——, A lower bound for linear forms in logarithms, to appear Acta Arith. **37** (1979)

[We 1] A. Weil, L'arithmétique sur les courbes algébriques, Acta Math. **52** (1928) pp. 281–315

[We 2] ——, Sur un théorème de Mordell, Bull. Sci. Math. **54** (1930) pp. 182–191

[We 3] ——, Sur les fonctions élliptiques p-adiques, C. R. Acad. Sci. Paris **203** (1936) pp. 22–24

[We 4] ——, Arithmétique et Géometrie sur les variétés algébriques, Actualités scientifiques et industrielles, No. 206, Hermann et Cie., Paris, 1935

[Zi] H. Zimmer, On the difference of the Weil height and the Néron-Tate height, Math. Z. **147** (1976) pp. 35–51

Index

Grundlehren der mathematischen Wissenschaften

A Series of Comprehensive Studies in Mathematics

A Selection

10. Schouten: Der Ricci-Kalkül
23. Pasch: Vorlesungen über neuere Geometrie
41. Steinitz: Vorlesungen über die Theorie der Polyeder
45. Alexandroff/Hopf: Topologie. Band 1
46. Nevanlinna: Eindeutige analytische Funktionen
57. Hamel: Theoretische Mechanik
63. Eichler: Quadratische Formen und orthogonale Gruppen
91. Prachar: Primzahlverteilung
102. Nevanlinna/Nevanlinna: Absolute Analysis
114. Mac Lane: Homology
127. Hermes: Enumerability, Decidability, Computability
131. Hirzebruch: Topological Methods in Algebraic Geometry
135. Handbook for Automatic Computation. Vol. 1/Part a: Rutishauser: Description of ALGOL 60
137. Handbook for Automatic Computation. Vol. 1/Part b: Grau/Hill/Langmaak: Translation of ALGOL 60
138. Hahn: Stability of Motion
139. Mathematische Hilfsmittel des Ingenieurs. 1. Teil
140. Mathematische Hilfsmittel des Ingenieurs. 2. Teil
141. Mathematische Hilfsmittel des Ingenieurs. 3. Teil
142. Mathematische Hilfsmittel des Ingenieurs. 4. Teil
143. Schur/Grunsky: Vorlesungen über Invariantentheorie
144. Weil: Basic Number Theory
145. Butzer/Berens: Semi-Groups of Operators and Approximation
146. Treves: Locally Convex Spaces and Linear Partial Differential Equations
147. Lamotke: Semisimpliziale algebraische Topologie
148. Chandrasekharan: Introduction to Analytic Number Theory
149. Sario/Oikawa: Capacity Functions
150. Iosifescu/Theodorescu: Random Processes and Learning
151. Mandl: Analytical Treatment of One-dimensional Markov Processes
152. Hewitt/Ross: Abstract Harmonic Analysis. Vol. 2: Structure and Analysis for Compact Groups. Analysis on Locally Compact Abelian Groups
153. Federer: Geometric Measure Theory
154. Singer: Bases in Banach Spaces I
155. Müller: Foundations of the Mathematical Theory of Electromagnetic Waves
156. van der Waerden: Mathematical Statistics
157. Prohorov/Rozanov: Probability Theory. Basic Concepts. Limit Theorems. Random Processes
158. Constantinescu/Cornea: Potential Theory on Harmonic Spaces
159. Köthe: Topological Vector Spaces I
160. Agrest/Maksimov: Theory of Incomplete Cylindrical Functions and their Applications
161. Bhatia/Szegö: Stability Theory of Dynamical Systems
162. Nevanlinna: Analytic Functions
163. Stoer/Witzgall: Convexity and Optimization in Finite Dimensions I
164. Sario/Nakai: Classification Theory of Riemann Surfaces
165. Mitrinović/Vasić: Analytic Inequalities
166. Grothendieck/Dieudonné: Eléments de Géometrie Algébrique I
167. Chandrasekharan: Arithmetical Functions
168. Palamodov: Linear Differential Operators with Constant Coefficients
169. Rademacher: Topics in Analytic Number Theory
170. Lions: Optimal Control of Systems Governed by Partial Differential Equations
171. Singer: Best Approximation in Normed Linear Spaces by Elements of Linear Subspaces
172. Bühlmann: Mathematical Methods in Risk Theory

173. Maeda/Maeda: Theory of Symmetric Lattices
174. Stiefel/Scheifele: Linear and Regular Celestial Mechanic. Perturbed Two-body Motion - Numerical Methods - Canonical Theory
175. Larsen: An Introduction to the Theory of Multipliers
176. Grauert/Remmert: Analytische Stellenalgebren
177. Flügge: Practical Quantum Mechanics I
178. Flügge: Practical Quantum Mechanics II
179. Giraud: Cohomologie non abélienne
180: Landkof: Foundations of Modern Potential Theory
181. Lions/Magenes: Non-Homogeneous Boundary Value Problems and Applications I
182. Lions/Magenes: Non-Homogeneous Boundary Value Problems and Applications II
183. Lions/Magenes: Non-Homogeneous Boundary Value Problems and Applications III
184. Rosenblatt: Markov Processes. Structure and Asymptotic Behavior
185. Rubinowicz: Sommerfeldsche Polynommethode
186. Handbook for Automatic Computation. Vol. 2. Wilkinson/Reinsch: Linear Algebra
187. Siegel/Moser: Lectures on Celestial Mechanics
188. Warner: Harmonic Analysis on Semi-Simple Lie Groups I
189. Warner: Harmonic Analysis on Semi-Simple Lie Groups II
190. Faith: Algebra: Rings, Modules, and Categories I
191. Faith: Algebra II, Ring Theory
192. Mal'cev: Algebraic Systems
193. Pólya/Szegö: Problems and Theorems in Analysis I
194. Igusa: Theta Functions
195. Berberian: Baer*-Rings
196. Athreya/Ney: Branching Processes
197. Benz: Vorlesungen über Geometrie der Algebren
198. Gaal: Linear Analysis and Representation Theory
199. Nitsche: Vorlesungen über Minimalflächen
200. Dold: Lectures on Algebraic Topology
201. Beck: Continuous Flows in the Plane
202. Schmetterer: Introduction to Mathematical Statistics
203. Schoeneberg: Elliptic Modular Functions
204. Popov: Hyperstability of Control Systems
205. Nikol'skii: Approximation of Functions of Several Variables and Imbedding Theorems
206. André: Homologie des Algèbres Commutatives
207. Donoghue: Monotone Matrix Functions and Analytic Continuation
208. Lacey: The Isometric Theory of Classical Banach Spaces
209. Ringel: Map Color Theorem
210. Gihman/Skorohod: The Theory of Stochastic Processes I
211. Comfort/Negrepontis: The Theory of Ultrafilters
212. Switzer: Algebraic Topology - Homotopy and Homology
213. Shafarevich: Basic Algebraic Geometry
214. van der Waerden: Group Theory and Quantum Mechanics
215. Schaefer: Banach Lattices and Positive Operators
216. Pólya/Szegö: Problems and Theorems in Analysis II
217. Stenström: Rings of Quotients
218. Gihman/Skorohod: The Theory of Stochastic Processes II
219. Duvaut/Lions: Inequalities in Mechanics and Physics
220. Kirillov: Elements of the Theory of Representations
221. Mumford: Algebraic Geometry I: Complex Projective Varieties
222. Lang: Introduction to Modular Forms
223. Bergh/Löfström: Interpolation Spaces. An Introduction
224. Gilbarg/Trudinger: Elliptic Partial Differential Equations of Second Order
225. Schütte: Proof Theory
226. Karoubi: K-Theory. An Introduction
227. Grauert/Remmert: Theorie der Steinschen Räume
228. Segal/Kunze: Integrals and Operators